FIRST CINVESTAV-UNAM SYMPOSIUM ON HIGH ENERGY PHYSICS

FIRST CINVESTAV-UNAM SYMPOSIUM ON HIGH ENERGY PHYSICS

Dedicated to the Memory of Augusto García

México City, Mexico
30 November - 1 December 2009

Cinvestav UNAM

EDITORS

Juan Carlos D'Olivo
Alejandro Frank
UNAM, Mexico

Ricardo López-Fernández
Miguel Ángel Pérez
CINVESTAV, Mexico

SPONSORING ORGANIZATIONS
Universidad Nacional Autónoma de México (UNAM)
CINVESTAV
CONACyT

AIP American Institute of Physics

Melville, New York, 2010
AIP CONFERENCE PROCEEDINGS ■ 1259

Editors

Juan Carlos D'Olivo
Alejandro Frank
Instituto de Ciencias Nucleares
UNAM
Apdo. Postal 70-543
04510 Mexico, D. F.
Mexico
E-mail: dolivo@nucleares.unam.mx
frank@nucleares.unam.mx

Ricardo López-Fernández
Miguel Ángel Pérez
Depto. Física, CINVESTAV, Apdo. Postal
14-740, 07000 México, D. F.
Mexico
E-mail: lopezr@fis.cinvestav.mx
mperez@fis.cinvestav.mx

L.C. Catalog Card No. 2010908083

ISBN 978-0-7354-0808-1

ISSN 0094-243X
Printed in the United States of America

CONTENTS

THEORY

Prologue

Augusto García, Professor of Physics at the Centro de Investigación y Estudios Avanzados (Cinvestav) del I.P.N., passed away in Mexico City on August 17, 2009. The present volume is a scientific homage to Augusto García put together by his friends, colleagues and former students. A collaboration agreement between Cinvestav and the National University of Mexico (Universidad Nacional Autónoma de México, UNAM) was signed in early 2009 and after Augusto's death both institutions promoted as a starting point of this agreement the celebration of a Symposium in honor of our friend. The First Cinvestav-UNAM Symposium on High Energy Physics was held in Mexico City on November 30, 2009 (Cinvestav, José Adem Auditorium, Zacatenco) and December 1, 2009 (UNAM, Marcos Moshinsky Auditorium, Instituto de Ciencias Nucleares, Ciudad Universitaria).

Augusto did his undergraduate studies at the Science Faculty of UNAM, got his Ph.D. degree in physics at the Fermi Institute of the University of Chicago and joined the Physics Department of Cinvestav in 1972. Augusto's research work was mainly concerned with electroweak phenomenology. However, Augusto's interests in physics were uncommonly far-ranging and was also concerned with the creation of new research groups in many Mexican institutions. In particular, he was instrumental in the creation of research groups both at UNAM and Cinvestav in two new lines of research in Mexico: experimental high energy physics and medical physics. In this respect, for many of us Augusto was a model physicist and an emblematic researcher to start the series of Cinvestav-UNAM Symposia on high energy physics.

We have organized the contributions to these proceedings in the following sections:

1. Biographical notes
2. Experiments and accelerators
3. Phenomenology
4. Theory

We are deeply grateful to the following members of Cinvestav and UNAM staff whose help made possible to have an excellent memorial symposium: Gabriela Frías, Hortensia Segura, Patricia Villar and Daniel Pérez. We are also indebted to the following Cinvestav and UNAM authorities for the financial support that made possible the celebration of the memorial symposium and the publication of the present proceedings: Arnulfo Albores, Carlos Aramburu de la Hoz and Isaac Hernández Calderón.

Juan Carlos D'Olivo
Alejandro Frank
Ricardo López-Fernández
Miguel A. Pérez

Augusto García González, 1942–2009

BIOGRAPHICAL NOTES

Augusto García: Colleague and Friend

P. Kielanowski

Departamento de Física, Centro de Investigación y de Estudios Avanzados del IPN, 07360 México D.F., Mexico

Abstract. In this paper I will try to remember the life of Augusto García, his achievements and contributions.

It is very difficult to write about Augusto García, who such a short time ago was with us. Now our memories are still vivid and have not disappeared in the shadow of the past, so I will try to recall Augusto as he really was and how we saw him.

EARLY LIFE, EDUCATION AND PROFESSIONAL CAREER

Augusto García González was born on September 12, 1942 in Mexico City and died on August 16, 2009. After graduation from high school in 1960, he entered the Facultad de Ciencias (Faculty of Science) of UNAM [1]. He graduated in 1965 as a "Físico" (Physicist) – equivalent of a Bachelor of Science degree. Upon matriculation he decided to become a scientist and chose the field of particle physics which was much better developed outside of Mexico. His choice was the University of Chicago where he moved with his wife and young son, Enrique. In 1967 he received the Masters of Science and 1971 completed his Ph.D. under the supervision of Reinhard Oehme. The topic of his research was the phenomenological analysis of the semileptonic hyperon decays. During his stay in Chicago his second son, Augusto, was born.

After receiving Ph.D. Augusto returned to Mexico and joined the Department of Physics of ESFM [2] of the National Polytechnic Institute. The following year, in 1972, Augusto moved to the Department of Physics at the Centro de Investigación y de Estudios Avanzados – Cinvestav[3] of the National Polytechnic Institute, where he remained until his death. During the period 1979–1985 he was also a part time professor at the ESFM with which he always kept close scientific contacts.

Augusto García joined Cinvestav at the same time as Arnulfo Zepeda and together they were two principal builders of the high energy physics group, which is one of the three most important research groups at the Department of Physics of Cinvestav.

---

[1] Universidad Nacional Autonóma de México.

[2] Escuela Superior de Física y Matemáticas.

[3] Center for Research and Advanced Studies

CP1259, *First CINVESTAV-UNAM Symposium on High Energy Physics*
edited by J. C. D'Olivo, R. López-Fernández, and M. Á. Pérez

CHICAGO TIMES: M.SC AND PH.D.

In 1966 Augusto García was enrolled as student in the Department of Physics at the University of Chicago and in 1967 received the M.Sc. degree and immediately entered into the Ph.D. program. His adviser was Reinhard Oehme[4], an outstanding physicist of German origin. The subject of his thesis was the precise analysis of the semileptonic decay of hyperons. The semileptonic decay of hyperons is described by the matrix elements of the vector and axial vector currents whose charges (or time components) form octets of the $SU(3)$ group. Thus the $SU(3)$ symmetry predicts definite and precise relations between the rates and asymmetries for the decays of various hyperons. The motivation for this research was the phenomenological analysis of the $SU(3)$ relations and the comparison of these relations with the existing experimental data. The first paper by Augusto had the title "Beta Decay of Hyperons" and was published in *Physical Review* in 1971. This paper, relatively short, dealt mostly with the analysis of various observables for the Λ beta decay. It was almost immediately followed by a more extensive and more detailed paper "Analysis of Spin Correlations in the β Decay of the Λ Hyperon" also published in the *Physical Review*. These two papers, were the cornerstone of a large part of the scientific effort of Augusto García. In Figs. 1 and 2 we include a copy of the initial parts of these papers.

AUGUSTO THE RESEARCH PHYSICIST

For Augusto García physics was not only his profession, it was also one of his passions. Let me start this section with two quotations of Augusto from his lecture *Critical view of the concepts of quantum mechanics* given at ESFM of the National Polytechnic Institute in 2005. The first quote is about the basic notions of physics

> ...I would like to make clear what is my attitude. This is physics. If you ask a physicist what is time, the answer is, it is what is measured with a clock, all the rest is metaphysics. What is space, and the answer is, it is what you measure with a ruler and everything else is metaphysics, or mathematics...

The second quote is about the basic principle of any scientific research, taking as an example the difference between the Copernican and Ptolomean theories of the solar system

> ...the essential point is the order of planets. Not the details of their motion, but their order. The two theories predict that if Venus is also a round object, like Earth, Moon or Sun, then Venus should also have phases, like the Moon. But the order of the phases of Venus in the Copernicus theory is opposite to the order in the Ptolemy theory. The observations demonstrated that the phases of Venus follow the theory of Copernicus. The conclusion is: one cannot say

[4] The adviser for the Oeheme's Ph.D. thesis was Werner Heisenberg. One can say that Augusto was scientifically a "grandson" of Heisenberg.

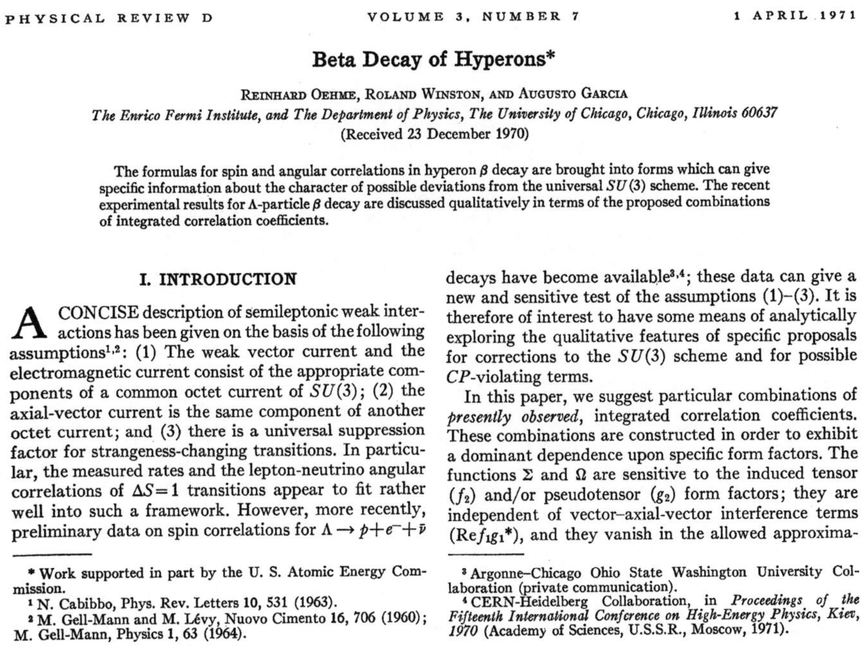

PHYSICAL REVIEW D VOLUME 3, NUMBER 7 1 APRIL 1971

Beta Decay of Hyperons*

REINHARD OEHME, ROLAND WINSTON, AND AUGUSTO GARCIA
The Enrico Fermi Institute, and The Department of Physics, The University of Chicago, Chicago, Illinois 60637
(Received 23 December 1970)

The formulas for spin and angular correlations in hyperon β decay are brought into forms which can give specific information about the character of possible deviations from the universal $SU(3)$ scheme. The recent experimental results for Λ-particle β decay are discussed qualitatively in terms of the proposed combinations of integrated correlation coefficients.

I. INTRODUCTION

A CONCISE description of semileptonic weak interactions has been given on the basis of the following assumptions[1,2]: (1) The weak vector current and the electromagnetic current consist of the appropriate components of a common octet current of $SU(3)$; (2) the axial-vector current is the same component of another octet current; and (3) there is a universal suppression factor for strangeness-changing transitions. In particular, the measured rates and the lepton-neutrino angular correlations of $\Delta S=1$ transitions appear to fit rather well into such a framework. However, more recently, preliminary data on spin correlations for $\Lambda \rightarrow p+e^-+\bar{\nu}$ decays have become available[3,4]; these data can give a new and sensitive test of the assumptions (1)–(3). It is therefore of interest to have some means of analytically exploring the qualitative features of specific proposals for corrections to the $SU(3)$ scheme and for possible CP-violating terms.

In this paper, we suggest particular combinations of *presently observed*, integrated correlation coefficients. These combinations are constructed in order to exhibit a dominant dependence upon specific form factors. The functions Σ and Ω are sensitive to the induced tensor (f_2) and/or pseudotensor (g_2) form factors; they are independent of vector–axial-vector interference terms ($\mathrm{Re} f_1 g_1^*$), and they vanish in the allowed approxima-

* Work supported in part by the U. S. Atomic Energy Commission.
[1] N. Cabibbo, Phys. Rev. Letters 10, 531 (1963).
[2] M. Gell-Mann and M. Lévy, Nuovo Cimento 16, 706 (1960); M. Gell-Mann, Physics 1, 63 (1964).
[3] Argonne–Chicago Ohio State Washington University Collaboration (private communication).
[4] CERN-Heidelberg Collaboration, in *Proceedings of the Fifteenth International Conference on High-Energy Physics, Kiev, 1970* (Academy of Sciences, U.S.S.R., Moscow, 1971).

FIGURE 1. The first page of the first paper by Augusto García. The paper has three authors, Reinhard Oehme was García's Ph.D. adviser.

> that Copernicus is right, and the only thing that one knows is that Ptolemy is wrong. This is the principle of falsification...

These two basic principles

- Measurability
- Falsification

shaped the topics of his research and the way how he did it.

The principal lines of research of Augusto García and the approximate dates when they were started are the following:

1. Semileptonic hyperon decay, 1966.
2. Radiative corrections, 1978.
3. Left-right symmetry, 1985.
4. A priori mixing and mirror matter, 1997.

The beta decay of hyperons was the subject of Augusto's Ph.D. and it was suggested by Reinhard Oehme. This topic remained an important subject throughout his career. Also in his later years, when Augusto was an independent scientist, he worked on semileptonic decays. The method that Augusto proposed and developed was based on

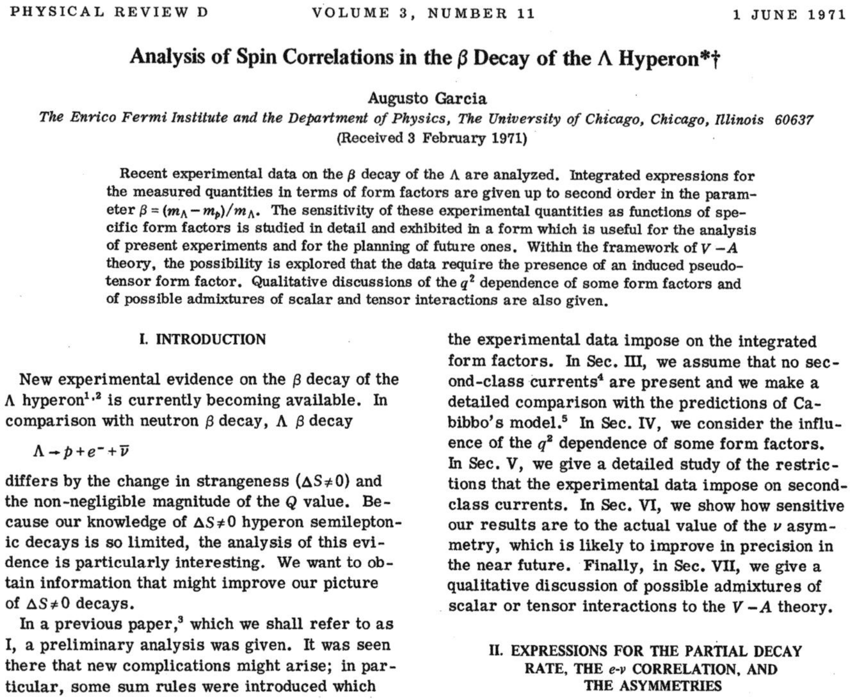

PHYSICAL REVIEW D VOLUME 3, NUMBER 11 1 JUNE 1971

Analysis of Spin Correlations in the β Decay of the Λ Hyperon*†

Augusto Garcia

The Enrico Fermi Institute and the Department of Physics, The University of Chicago, Chicago, Illinois 60637

(Received 3 February 1971)

Recent experimental data on the β decay of the Λ are analyzed. Integrated expressions for the measured quantities in terms of form factors are given up to second order in the parameter $\beta = (m_\Lambda - m_p)/m_\Lambda$. The sensitivity of these experimental quantities as functions of specific form factors is studied in detail and exhibited in a form which is useful for the analysis of present experiments and for the planning of future ones. Within the framework of $V-A$ theory, the possibility is explored that the data require the presence of an induced pseudotensor form factor. Qualitative discussions of the q^2 dependence of some form factors and of possible admixtures of scalar and tensor interactions are also given.

I. INTRODUCTION

New experimental evidence on the β decay of the Λ hyperon[1,2] is currently becoming available. In comparison with neutron β decay, Λ β decay

$$\Lambda \to p + e^- + \bar{\nu}$$

differs by the change in strangeness ($\Delta S \neq 0$) and the non-negligible magnitude of the Q value. Because our knowledge of $\Delta S \neq 0$ hyperon semileptonic decays is so limited, the analysis of this evidence is particularly interesting. We want to obtain information that might improve our picture of $\Delta S \neq 0$ decays.

In a previous paper,[3] which we shall refer to as I, a preliminary analysis was given. It was seen there that new complications might arise; in particular, some sum rules were introduced which the experimental data impose on the integrated form factors. In Sec. III, we assume that no second-class currents[4] are present and we make a detailed comparison with the predictions of Cabibbo's model.[5] In Sec. IV, we consider the influence of the q^2 dependence of some form factors. In Sec. V, we give a detailed study of the restrictions that the experimental data impose on second-class currents. In Sec. VI, we show how sensitive our results are to the actual value of the ν asymmetry, which is likely to improve in precision in the near future. Finally, in Sec. VII, we give a qualitative discussion of possible admixtures of scalar or tensor interactions to the $V-A$ theory.

II. EXPRESSIONS FOR THE PARTIAL DECAY RATE, THE e-ν CORRELATION, AND THE ASYMMETRIES

FIGURE 2. The beginning page of the second paper by Augusto García.

the idea that the comparison between the theory and the experimental results should be done for the observables like rates or asymmetries and not for the form factors. Such a method is much cleaner and gives a deeper insight on the interpretation of the experimental results.

The summary of the results of this research is contained in the monograph "The Beta Decay of Hyperons", published as a volume in the series *Lecture Notes in Physics* by the Springer Verlag in 1986. The monograph, which contains detailed study of the beta decay of hyperons, was highly valued and extensively used by experimental physicists working in the field. It contains a detailed description of how the experimental data need to be tested. The result of his analysis was to show, that $SU(3)$ symmetry is broken in beta decays of hyperons.

Augusto's next subject, *radiative corrections*, is a logical consequence of the above result: one needs radiative corrections for an accurate explanation of the beta decay data of hyperons. The improvement in the statistics of the experiments increased the precision of the experimental data. The theoretical predictions contain approximations and the size of the radiative corrections became larger than the statistical errors of the data. Augusto undertook the task of the calculation of the radiative corrections for the beta decay of hyperons and he did it in his own unique manner: He calculated radiative corrections for

FIGURE 3. The cover of the monograph "The Beta Decay of Hyperons".

the experimental observables and in such a way greatly improved the precision of these tests. Augusto's radiative corrections were included in all the tests that were published after 1978.

The next subject of Augusto's research studies is the *left right symmetry in the standard model.* Its origin was his quest for a larger symmetry and the experimental evidence that might support it. One can say that the first paper on the subject in 1985 was a preparatory one. In this paper Augusto and his co-workers consider a test for the presence of right handed currents in semileptonic baryon decays. In the following paper in 1991 they consider the left-right symmetric extension of the standard model and obtain very precise limits on the parameters of the model. It is interesting to note that Augusto was looking for the evidence of the left-right symmetry in the neutron beta decay. This is a beautiful example how very precise low energy measurements of simple processes can be used to obtain bounds on extensions of the standard model. This work also shows Augusto's deep understanding of the semileptonic decays of hyperons.

The last topic that can clearly be distinguished in Augusto's research is the presence of the *a priori* mixings in the eigenstates of the physical states of particles. The precise explanation of the *a priori* mixings is given in the first paper on this subject:

> ...Because parity and strong flavors (strangeness, charm, etc.) are violated in nature, the physical (mass eigenstates) hadrons cannot be either parity

or flavor eigenstates, i.e., the former must be admixtures of the latter. It is generally believed that the breaking of flavor global groups is caused by the mass differences of hadrons, but in such a way that parity and all flavors are conserved, i.e., the mass operator of hadrons giving rise to such breakings does not contain a piece that violates parity and flavor. The flavor and parity mixings in physical hadrons are attributed to the perturbative intervention of $W_{\mu}^{\pm}$ and Z^0 (parity mixing only). And, precisely because such intervention is perturbative, such mixings can appear only in higher orders of perturbation theory; thus, such mixings appear, so to speak, *a posteriori*.

However, the possibility that the mass operator of hadrons does contain a (necessarily) very small piece that is flavor and parity violating is not excluded by any fundamental principle. If such a piece does exist, then, the parity and flavor admixtures in hadrons must come *a priori*, in a non-perturbative way...

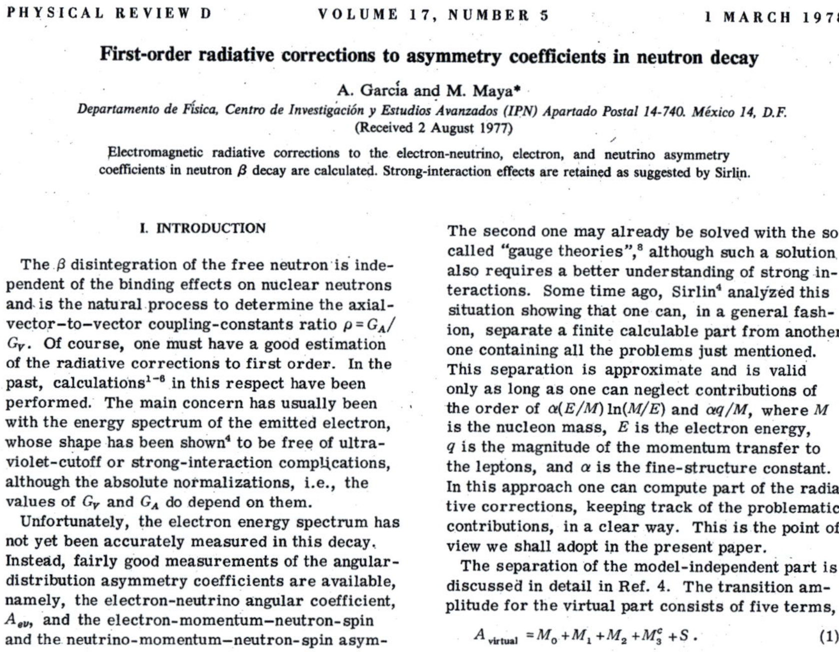

PHYSICAL REVIEW D VOLUME 17, NUMBER 5 1 MARCH 1978

First-order radiative corrections to asymmetry coefficients in neutron decay

A. García and M. Maya*

Departamento de Física, Centro de Investigación y Estudios Avanzados (IPN) Apartado Postal 14-740. México 14, D.F.

(Received 2 August 1977)

Electromagnetic radiative corrections to the electron-neutrino, electron, and neutrino asymmetry coefficients in neutron β decay are calculated. Strong-interaction effects are retained as suggested by Sirlin.

I. INTRODUCTION

The β disintegration of the free neutron is independent of the binding effects on nuclear neutrons and is the natural process to determine the axial-vector–to–vector coupling-constants ratio $\rho = G_A/G_V$. Of course, one must have a good estimation of the radiative corrections to first order. In the past, calculations[1-6] in this respect have been performed. The main concern has usually been with the energy spectrum of the emitted electron, whose shape has been shown[4] to be free of ultraviolet-cutoff or strong-interaction complications, although the absolute normalizations, i.e., the values of G_V and G_A do depend on them.

Unfortunately, the electron energy spectrum has not yet been accurately measured in this decay. Instead, fairly good measurements of the angular-distribution asymmetry coefficients are available, namely, the electron-neutrino angular coefficient, $A_{e\nu}$, and the electron-momentum–neutron-spin and the neutrino-momentum–neutron-spin asym- The second one may already be solved with the so called "gauge theories",[8] although such a solution also requires a better understanding of strong interactions. Some time ago, Sirlin[4] analyzed this situation showing that one can, in a general fashion, separate a finite calculable part from another one containing all the problems just mentioned. This separation is approximate and is valid only as long as one can neglect contributions of the order of $\alpha(E/M)\ln(M/E)$ and $\alpha q/M$, where M is the nucleon mass, E is the electron energy, q is the magnitude of the momentum transfer to the leptons, and α is the fine-structure constant. In this approach one can compute part of the radiative corrections, keeping track of the problematic contributions, in a clear way. This is the point of view we shall adopt in the present paper.

The separation of the model-independent part is discussed in detail in Ref. 4. The transition amplitude for the virtual part consists of five terms,

$$A_{\text{virtual}} = M_0 + M_1 + M_2 + M_3^c + S\,. \quad (1)$$

FIGURE 4. The initial page of the first paper by Augusto García on radiative corrections.

The research on the *a priori* mixed eigenstates resulted in a series of papers, where Augusto with collaborators demonstrated the usefulness of the notion of the *a priori* mixtures in the analysis of the properties of baryons and mesons. In particular they applied the method to the discussion of such phenomena as weak radiative and non-leptonic decays of hadrons and $\Delta I = 1/2$ rule for mesons and hyperons. This research also led to the study of mirror baryons, where a stringent lower bound of 10^6 GeV for the mass of mirror partners was obtained.

PHYSICAL REVIEW D VOLUME 22, NUMBER 5 1 SEPTEMBER 1980

General form of the β energy spectrum in charged-hyperon β decay

A. García*

Departamento de Física, Centro de Investigación y Estudios Avanzados (IPN), Apartado Postal 14-740, México 14, D.F., México

S. R. Juárez W.

Escuela Superior de Física y Matemáticas (IPN), Apartado Postal 14-740, México 14, D.F., México

(Received 28 December 1979)

We have computed the radiative corrections to the β energy spectrum in the semileptonic decay of charged hyperons. Strong-interaction and intermediate-vector-boson effects on the electromagnetic corrections are retained in general by extending an approach that Sirlin introduced for neutron decay. Our result is quite accurate. It is valid up to corrections no larger than a few tenths of a percent. It is suitable for performing experimental analysis in a model-independent fashion.

INTRODUCTION

With the advent of hyperon beams, experiments that measure the energy spectrum of the electron or positron emitted in the β decay of charged hyperons can be performed. The analysis of such experiments will require a knowledge of radiative corrections. Our aim in the present paper is to obtain a theoretical expression for the energy spectrum of the electron or positron that includes the radiative corrections to first order in α, the decay. It consists of separating a finite model-independent part from the most general form of the radiative corrections within the $V-A$ theory, allowing for the effects of strong interactions and the presence of an intermediate vector boson with all generality. We shall follow this approach to handle the radiative corrections in the β decay of charged hyperons. We shall end with an expression for the energy spectrum of the β that is very accurate and can be used in performing unbiased experimental analysis.

FIGURE 5. The initial page of the frequently cited paper by Augusto García on radiative corrections.

Summarizing, it can be said that the most important research topic of Augusto García was the semileptonic decays and without any doubt one can say that Augusto was the leading world expert on this subject. The remaining fields (like radiative corrections) of his interest were related or stemmed from the semileptonic hyperon decays. Though Augusto's speciality was relatively narrow he had a deep knowledge of field theory and quantum mechanics which determined the quality of his research. This can be seen in his papers and is also documented by his teaching.

AUGUSTO THE TEACHER

The main professional activity of Augusto García was research – the quest for new knowledge and methods of investigation. The other area, where he left a very significant mark was teaching. He regularly was giving courses on various subjects of theoretical physics (classical and quantum mechanics, electrodynamics). He was also Ph.D. and M.Sc. adviser of many students, who (together with their students) are now working in various scientific institutions all over Mexico. One can say that there are some 60 active physicists, who were formed by Augusto or his students. This means that his impact on the Mexican scientific life is very significant and will remain for a long time to come. His way of thinking, method of work and passion for physics were passed to subsequent generations.

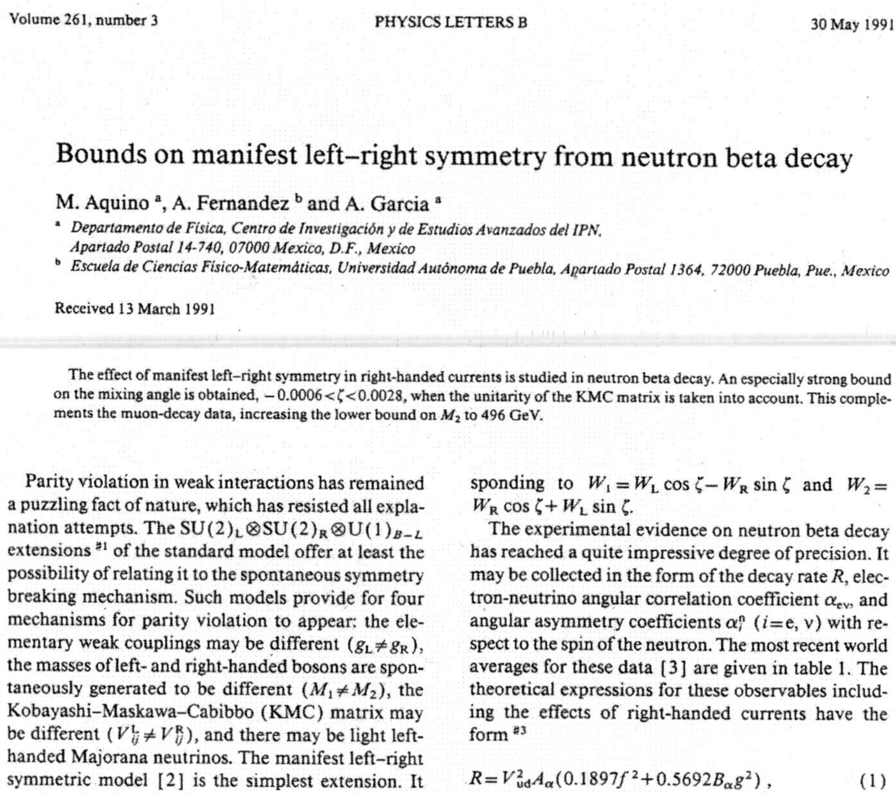

Volume 261, number 3 PHYSICS LETTERS B 30 May 1991

Bounds on manifest left–right symmetry from neutron beta decay

M. Aquino [a], A. Fernandez [b] and A. Garcia [a]

[a] *Departamento de Física, Centro de Investigación y de Estudios Avanzados del IPN, Apartado Postal 14-740, 07000 Mexico, D.F., Mexico*

[b] *Escuela de Ciencias Físico-Matemáticas, Universidad Autónoma de Puebla, Apartado Postal 1364, 72000 Puebla, Pue., Mexico*

Received 13 March 1991

The effect of manifest left–right symmetry in right-handed currents is studied in neutron beta decay. An especially strong bound on the mixing angle is obtained, $-0.0006<\zeta<0.0028$, when the unitarity of the KMC matrix is taken into account. This complements the muon-decay data, increasing the lower bound on M_2 to 496 GeV.

Parity violation in weak interactions has remained a puzzling fact of nature, which has resisted all explanation attempts. The $SU(2)_L\otimes SU(2)_R\otimes U(1)_{B-L}$ extensions [#1] of the standard model offer at least the possibility of relating it to the spontaneous symmetry breaking mechanism. Such models provide for four mechanisms for parity violation to appear: the elementary weak couplings may be different ($g_L\neq g_R$), the masses of left- and right-handed bosons are spontaneously generated to be different ($M_1\neq M_2$), the Kobayashi–Maskawa–Cabibbo (KMC) matrix may be different ($V^L_{ij}\neq V^R_{ij}$), and there may be light left-handed Majorana neutrinos. The manifest left–right symmetric model [2] is the simplest extension. It sponding to $W_1=W_L\cos\zeta-W_R\sin\zeta$ and $W_2=W_R\cos\zeta+W_L\sin\zeta$.

The experimental evidence on neutron beta decay has reached a quite impressive degree of precision. It may be collected in the form of the decay rate R, electron-neutrino angular correlation coefficient $\alpha_{e\nu}$, and angular asymmetry coefficients α_i^n (i=e, ν) with respect to the spin of the neutron. The most recent world averages for these data [3] are given in table 1. The theoretical expressions for these observables including the effects of right-handed currents have the form [#3]

$$R=V_{ud}^2A_\alpha(0.1897f^2+0.5692B_\alpha g^2)\,, \quad (1)$$

FIGURE 6. The initial page of the main paper by Augusto García on the analysis of the left-right symmetric models.

The list of Augusto's students is the following[5]

M.Sc.

M. Maya (1974)
R. Huerta (1978)
M. Avila (1983)
A. Morelos (1985)
D. Tun (1986)
P. Ferreira(1987)
M. Aquino (1987)
E. Esparza (1991)
R. Flores (1994)

[5] For the Ph.D. students I also give the number of M.Sc. and Ph.D. students graduated by them.

J.L. García Luna (1997)
M. Neri (2002) (with A. Martinéz)

Ph.D.

J.M. Rivera (1980) - 3 M.Sc. and 1 Ph.D.
A. Queijeiro (1982)
R. Juárez (1983) - 6 M.Sc. and 1 Ph.D.
M. Avila (1987)
D. Tun (1988) - 3 M.Sc. and 1 Ph.D.
A. Martínez (1988)- 6 M.Sc. and 1 Ph.D.
A. Morelos (1992) (with J. Lach) - 4 M.Sc.
M. Maya (1994)
O. Miranda (1997) (with M. Maya) - 5 M.Sc. and 2 Ph.D.
E. Esparza (1997) (with R. Huson)
R. Flores (1997) - 3 M.Sc.
J.L. García L. (2002) - 3 M.Sc. and 1 Ph.D.

Augusto was a very good teacher and adviser, because his attitude towards students was serious and demanding. He wanted the students to learn what he knew. He was often very persistent in his questioning during seminars – what the student knew what was his method, why he was applying a certain method, etc. He wanted to be sure that a student had as deep knowledge as possible. I think that together with the quality of his research this is the reason why he formed so many outstanding physicists.

ORGANIZER

The scientific achievements of Augusto García were recognized very early and this brought him into various administration tasks. It must be said that he never was interested in an administrative career, so his participation in these tasks was rather a necessary community service and this was the reason why he was undertaking them. I think that Augusto's role in these tasks was very important and productive, because he had always a broader view and never was indifferent. Also, after five years at the University of Chicago, he knew what are important elements of a good scientific institution, what are requirements that produce a productive and successful career of both professor or a student.

I will list here the posts that Augusto held in hist lifetime without going into details or trying to evaluate their importance.

Committee for scholarships of CONACyT [6], (1972-1976, 1982-1990).
Committee of special programs of CONACyT (1984-1988).
Committee of assistance for postgraduate studies of CONACyT (1984-1988).
Committee of scientific projects of CONACyT (1988-90) (President in 1990).

---

[6] CONACyT stands for Consejo Nacional de Ciencia y Tecnología – National Council of Science and Technology (Mexico).

Committee of support for active research groups, Academy of Scientific Research, (1988-89).
Selection committee for new members, Academy of Scientific Research, (1989-1993).
Council for the Ph.D. program at the University of Puebla from 1982.
Evaluation committee of the fourfold support program for the Institute of Physics at the University of Guanajuato.
Advisory committee in the Institute of Physics at UNAM.
Academic Committee for the Ph.D. program, DIE-Cinvestav, from 1987.
Commission for promotion and stimuli of the investigators, Cinvestav (COPEI),
    Founding member (1980-1987),
    President (1986-1987).
Committee of scientific programs of CONACyT, (1994-1996).
Committee of the Mexican National Prize at the office of the President of Mexico (1997).
Member of Academic Consultive Council, Cinvestav; elected as the representative for the area of exact sciences (1997-1999).
Member of the Academic council of the Department of Physics at the Iberoamerican University (2004, 2005).
Academic coordinator, Department of Physics, Cinvestav (1976-1977).
Chairman of the Department of Physics, Cinvestav (1980-1984).

The list is impressive and one can see that the amount time, effort and thinking Augusto had to dedicate to fulfill all these duties was significant.

Apart from the "official" duties mentioned above Augusto played very important role in the development of the high energy physics group at Cinvestav. He joined the Department of Physics at Cinvestav together with Arnulfo Zepeda in 1972, when particle physics was rather scarcely represented. It took some 15 years and a lot of patience and work of three people: Augusto García, Arnulfo Zepeda and later Miguel Angel Pérez to form first the strong theoretical particle physics group (all three are theoreticians) and later the experimental particle physics group in Cinvestav, which further included also the medical physics. The role of Augusto García as well as of the two others was extremely important in this formation and their great vision and efforts have to be acknowledged.

AUGUSTO IN PRIVATE: FAMILY AND PASTIME

The successful professional life is frequently accompanied by a very harmonious personal life and this was the case of Augusto García. Augusto with his wife Quety (Enriqueta) and three sons (Enrique, Augusto and Pablo) formed a wonderful family, where love and respect abounded. Each of their sons received a university degree, Enrique is the medical doctor, Augusto-son is the physicist and Pablo – the composer. They choose their field by themselves, but what is important they could discuss at home their decisions and have excellent examples and good advice. All three are independent, but they

FIGURE 7. Some members of the high energy physics group around 1986 [7].

have always maintained very close contact with their parents.

Augusto, despite his many assignments, spent a lot of time with his family, frequently going together to concerts, cinema or for vacation. I think that they enjoyed their time together; Augusto's and Quety's hobbies were also attractive for their sons. Of course, as time passed, the sons became independent and they were spending less time together, but the familial bond was always very strong and endured throughout his lifetime.

Augusto was a great lover of music. He did not play an instrument, but listened to music frequently and attended many concerts with Quety. He was particularly partial to classical singing whether in operas, operettas or zarzuelas. Augusto played for me one of the most beautiful pieces of music, the duet "Là ci darem la mano" from Mozart's "Don Giovanni". I was impressed.

EPILOGUE

Augusto García is no more with us. They say that there is no one that is not replaceable. It is probably true. For our future students Augusto García will be only a name and

[7] The names of the people in the picture from right to left are: A. García, M. Socolovsky, E. Cantoral, R. Huerta, P. Kielanowski, M.A. Pérez, A. Zepeda, J.L. Lucio.

not a whole, complex person. But what will remain through his achievements and his students is that Augusto played a very important role in life and development of the Department of Physics at Cinvestav and left very a significant mark on the Mexican scientific landscape.

ACKNOWLEDGMENTS

During the preparation of this paper I greatly benefited from the advice and help of the following people: Darlene Bohm, S. Rebeca Juárez Wysozka, Gabriel López Castro and Miguel Angel Pérez Angón.

PUBLICATIONS OF AUGUSTO GARCÍA

Papers in journals

1. *Hyperon Beta Decay*, Phys. Rev. D **3**, 1618-1621 (1971) (with R. Oehme and R. Winston).
2. *Analysis of Spin Correlations in the Beta Decay of the Lambda Hyperon*, Phys. Rev. D **3**, 2638-2648 (1971).
3. *Current-Current Commutators, Slices of the Light Cone and the Value of the Sigma Term*, Phys. Rev. D **8**, 1820-1827 (1973) (with G. Cocho and J.J. Giambiagi).
4. *Test of Cabibbo's Model in Hyperon Beta Decay*, Phys. Rev. D **9**, 177-180 (1974).
5. *Sensitivity of a Recent Test of Cabibbo's Model to the Present Experimental Situation*, Nuvo Cimento **22**, 342-344 (1974).
6. *Cabibbo's Model with First Order $SU(3)$-Symmetry Breaking Corrections and Hyperon Beta Decay*, Phys. Rev. D **10**, 884-890 (1974).
7. *Restrictions on Axial-Vector Second Class Currents from Baryon Semileptonic Decay Data*, Phys. Rev. D **10**, 2839-2848 (1974).
8. *Implications of Recent Data on $\Sigma^- \to ne\nu$ for the Cabibbo Model*, Phys. Rev. Lett. **35**, 467-469 (1975) (with E.C. Swallow).
9. *Weak Magnetism and Pseudotensor form Factor in $\Sigma^- \to ne\nu$*, Phys. Rev. D **12**, 2692-2699 (1975).
10. *Spectrum Generating $SU(3)$ and the Semileptonic Decay of Hyperons*, Phys. Rev. D **15**, 689-695 (1977) (with A. Bohm, J. Nielsson and R. Teese).
11. *First-order Radiative Corrections to Asymmetry Coefficients in Neutron Decay*, Phys. Rev. D **17**, 1376-1380 (1978) (with M. Maya).
12. *Model Independent form of Certain Observables in Neutron Decay*, Phys. Lett. **73B**, 299-301 (1978).
13. *General Applicability of Certain Test of the $V-A$ Theory of Neutron Decay*, Pis'ma Zh. Eksp. Teor. Fis. **27**, 541-544 (1978) (English Translation: JETP Lett. **27**, 510 (1978).

14. *General Form of the Beta Energy Spectrum in Charged Hyperon Beta Decay*, Phys. Rev. D **22**, 1132-1141 (1980) (with S.R. Juárez).
15. *Radiative Corrections to $\tau \to \pi\nu$ Decay*, Nucl. Phys. **B189**, 500-508 (1981) (with J.M. Rivera).
16. *General Treatment of Electromagnetic Corrections to the Leptonic Decays of PseudoScalar Mesons*, Phys. Rev. D **23**, 1562-1566 (1981) (with A. Queijeiro).
17. *Model Independent Radiative Corrections to $M^{\pm}_{l3}$ Decays*, Phys. Rev. D **23**, 2603-2609 (1981) (with M. Maya).
18. *The Radiative Corrections Independent Form of Certain Observables in Semileptonic Decays of Hyperons*, Phys. Lett. **105B**, 224-226 (1981).
19. *Electromagnetic Corrections to the Semileptonic Decays of Polarized Neutral and Charged Hyperons*, Phys. Rev. D **25**, 1348-1354 (1982).
20. *Critical Analysis of Hyperon Semileptonic Decays*, Phys. Rev. D **25**, 1451-1454 (1982) (with P. Kielanowski).
21. *On the Experimental Determination of the Leptonic Decay Constant of Pseudoscalar Mesons*, Phys. Rev. D **25**, 1985-1988 (1982).
22. *A Highly Significant Deviation from the Cabiboo Theory in Hyperon Leptonic Decay*, Phys. Lett. **110B**, 498-500 (1982) (with P. Kielanowski).
23. *The Electron Asymmetry in $\Sigma^- \to ne\nu$ and Second Class Currents*, Phys. Lett. **113B**, 272-274 (1982) (with A. Bohm and P. Kielanowski).
24. *The Cabiboo Theory and Hyperon Semileptonic Decays*, Phys. Rev. D **26**, 1090-1102 (1982) (with P. Kielanowski).
25. *Symmetry Breaking and Higher Representations in the Cabiboo Theory*, Phys. Rev. D **26**, 3100-3107 (1982) (with P. Kielanowski).
26. *Some Theorems on the Determination of Form Factors in Unpolarized Hyperon Semileptonic Decays*, Phys. Lett. **120B**, 214-216 (1983) (with P. Kielanowski).
27. *Two Theorems on the Unique Determination on Form Factors in Polarized Hyperons Semileptonic Decays*, Phys. Lett. **121B**, 424-428 (1983) (with P. Kielanowski).
28. *Fits ot the Hyperon-Semileptonic-Decay Data and Spectrum-Generating $SU(3)$. II*, Phys. Rev. D **27**, 180-187 (1983) (with A. Bohm, P. Magnollay and P. Kielanowski).
29. *Radiative Corrections to the Leptonic Decays of Pseudoscalar Mesons: Contributions of W and Z Mesons*, Phys. Rev. D **27**, 2101-2106 (1983) (with A. Queijeiro).
30. *Electromagnetic Corrections to Semileptonic Decays with a Polarized Emitted Hyperon*, Phys. Rev. D **28**, 1659-1663 (1983).
31. *Logarithmic Lepton-Mass Singularities in Baryon Decay*, Phys. Rev. D **28**, 2180-2190 (1983) (with S.R. Juárez).
32. *Consistency between two recent experiments on $\Sigma^- \to ne\nu$*, Phys. Rev. D **30**, 231-233 (1984) (with A. Bohm and P. Kielanowski).
33. *Model Independence of the Logarithmic Lepton-Mass Singularities in the Radiative Corrections to M_{l3} Decay*, Phys. Rev. D **30**, 991-995 (1984) (with S.R. Juárez).
34. *Test of Right Handed Currents in Baryon Semileptonic Decays*, Phys. Lett. **157B**, 317-320 (1985) (with R. Huerta, M. Maya and R. Pérez).

35. *The values of the $\Sigma^- \to ne\nu$ Decay Parameters and the Theory of Leptonic Decays*, Rev. Mex. Fís. *31*, 361-383 (1985) (with A. Bohm, R. Teese and M. Kmiecik).
36. *β energy spectrum in charged hyperon semileptonic decays*, Phys. Rev. D **35**, 232-238 (1987) (with A. Martínez and S.R. Juárez).
37. *Discrepancy between theory and Experiment for $\Lambda_c^+ \to \Lambda^0 e^+ \nu$: Possible Evidence for a fourth generation*, Phys. Rev. Lett. **59**, 864-867 (1987) (with R. Huerta and P. Kielanowski).
38. *Radiative corrections to $\tau \to V + \nu_\tau$ decay*, Phys. Rev. D **38**, 2218-2225 (1988) (with A. Queijeiro).
39. *Analytic expression for baryon semileptonic decay rate*, Phys. Rev. D **37**, 2537-2541 (1988) (with B. González and R. Huerta).
40. *Further contributions to radiative corrections to the β energy spectrum of charged-hyperon semileptonic decays*, Phys. Rev. D **38**, 2904-2905 (1988) (with S.R. Juárez and A. Martínez).
41. García, Huerta and Kielanowski. Reply, Phys. Rev. Lett. **61**, 2505-2505, (1988).
42. *Predictions for semileptonic decays of charm baryons. I. $SU(4)$-Symmetry Limit*, Phys. Rev. D **40**, 2944-2954 (1989) (with M. Avila-Aoki, R. Huerta and R. Pérez-Marcial).
43. *Predictions for semileptonic decays of charm Baryons. II. Nonrelativistic and MIT quark models* Phys. Rev. D **40**, 2955-2966 (1989) (with R. Pérez-Marcial, R. Huerta and M. Avila-Aoki).
44. *Radiative corrections to the Dalitz plot of semileptonic decays of charged and neutral hyperons*, Phys. Rev. D **40**, 2967-2979 (1989) (with D.M. Tun and R. Juárez W.).
45. Further contributions to radiative corrections to the beta energy spectrum in charged-hyperon semileptonic decays, Phys. Rev. D **43**, 282-283 (1991) (with S.R. Juárez and A. Martínez).
46. *Bounds on manifest left-right symmetry from neutron beta decay*, Phys. Lett. **261B**, 280-284 (1991) (with A. Fernández and M. Aquino).
47. *Momentum transfer contributions to the radiative corrections of the Dalitz plot of semileptonic decays of charged baryons with light or charm quarks*. Phys. Rev. D **44**, 3589-3598 (1991) (with D. Tun and S.R. Juárez).
48. *The dependence on the choice of a basic set of form factors of the Low amplitude for hyperon semileptonic decays*, Rev. Mex. Fís. **37**, 452-460 (1991) (with S.R. Juárez and A. Martínez).
49. *Value of the Kobayashi-Maskawa-Cabibbo matrix element V_{us} and flavor summetry breaking in hyperon semileptonic decays*, Phys. Rev. D **45**, 879-883 (1992) (with R. Huerta and P. Kielanowski).
50. *Semileptonic decays of charm baryons*, Phys. Rev. D **45**, 3266-3267 (1992) (with R. Huerta).
51. *Numerical calculation of radiative corrections to the energy spectrum of semileptonic decays of light-and charm-quark charged baryons*, Phys. Rev. D **46**, 322-336 (1992) (with A. Martínez and S.R. Juárez).

52. *Radiative corrections to the Dalitz plot of semileptonic decays of neutral baryons with light or charm quarks*, Phys. Rev. D **47**, 3984-3993 (1993) (with A. Martínez and D.M. Tun).
53. *Radiative corrections to the Dalitz plot of semileptonic decays of neutral baryons with light or charm quarks*, Phys. Rev. D **50**, 2325-2328 (1994) (with A. Martínez, D.M. Tun and G. Sánchez-Colón).
54. *Determination of the Kobayashi-Maskawa-Cabibbo matrix element V_{us} under various flavor-summetry breaking models in hyperon semileptonic decays*, Phys. Rev. D **54**, 6855-6860 (1996) (with R. Flores and G. Sánchez-Colón).
55. *A priori mixed baryons and weak radiative decays*, Rev. Mex. Fís. **43**, 232-239 (1997).
56. *Radiative corrections to the semileptonic Dalitz plot with angular correlation between polarized decaying and emitted hyperons* Phys. Rev. D **55**, 5702-5718 (1997) (with R. Flores, A. Martínez and J.J. Torres).
57. *A priori mixed hadrons, weak radiative and non-leptonic decays of hyperons*, Jour. of Phys. G **24**, 1207-1217(1998) (with R. Huerta and G. Sánchez-Colón).
58. *A detailed determination of the a priori mixing angles in non-leptonic decays of hyperons*, Jour. of Phys. G **25** 45-57(1999) (with R. Huerta and G. Sánchez-Colón).
59. *Hyperon non-leptonic decays, the $\Delta I = 1/2$ rule, and a priori mixed hadrons*, Rev. Mex. Fís. **45** 244-248 (1999) (with R. Huerta and G. Sánchez-Colón).
60. *On the universality of the a priori mixing angles in weak radiative decays of hyperons*, Jour. of Phys. G **25**, L1-L6 (1999) (with R. Huerta and G. Sánchez-Colón).
61. *Symmetry limit properties of a priori mixing amplitudes for non-leptonic and weak radiative decays of hyperons*, Jour. of Phys. G **26**, 1417-1422 (2000) (with R. Huerta y G. Sánchez-Colón).
62. *A priori mixing of mesons and the $\Delta I = 1/2$ rule in $K \to \pi\pi$*, Mod. Phys. Lett. **A15**, 1749-1754 (2000) (with R. Huerta and G. Sánchez-Colón).
63. *Radiative corrections to the semileptonic Dalitz plot with angular correlation between polarized decaying hyperons and charged emitted leptons*, Phys. Rev. D **63**, 014025, 1-17 (2001) (with R. Flores, A. Martínez and J.J. Torres).
64. *An Estimate of a lower bound on the masses of mirror baryons*, Phys. Lett. **B498**, 251-255 (2001) (with R. Huerta and G. Sánchez-Colón).
65. *Neutron beta decay and the current determination of V_{ud}*, Phys. Lett. **B500**, 66-74 (2001) (with J.L. García-Luna and G. López-Castro).
66. *Radiative corrections to the semileptonic Dalitz plot with angular correlation between polarized decaying hyperons and emitted hyperons: effects of the four-body region*, Phys. Rev. D **65**, 074002, 1-6 (2002) (with A. Martínez, J.J. Torres and R. Flores-Mendieta).
67. *Radiative corrections to all charge assignments of heavy quark baryon semileptonic decays*, Phys. Rev. D **66**, 074014, 1-9 (2002) (with A. Martínez, J.J. Torres and R. Flores-Mendieta).

68. *Improved predictions of the standard model and the detection of new physics in neutron beta decay*, Phys. Lett. **B546**, 247-252 (2002) (with J.L. García-Luna).
69. *Mirror matter mixtures in W-two body non-leptonic decays and the $\Delta I = 1/2$ rule*, J. Phys. G: Nucl. Part. Phys. **29** (2003) L15-L20 (with G. Sánchez-Colón and R. Huerta).
70. *Mirror matter admixtures and isospin breaking in the $\Delta I = 1/2$ rule in W-two body non-leptonic decays*, Int. J. Mod. Phys. **A 19**, 775-784 (2004) (with G. Sánchez-Colón and R. Huerta).
71. *Mirror matter Admixtures in $K_L \to \gamma\gamma$*, Mod. Phys. Lett. **A 19**, 3019-3029 (2004) (with G. Sánchez-Colón).
72. *Numerical precision radiative corrections to the Dalitz plot of baryon semileptonic decays including the spin-momentum correlation of the decaying and emitted baryons*, Phys. Rev. D **70**, 093012 1-17 (2004) (with J.J. Torres, R. Flores-Mendieta, M. Neri and A. Martínez).
73. *Numerical precision radiative corrections to the Dalitz plot of light and heavy quark unpolarized baryon semileptonic decays: The cases $\Xi^0 \to \Sigma^+ e\nu$ and $\Lambda_c^+ \to \Lambda e^+$*, Phys. Rev. D **71**, 034023 1-10 (2005) (with R. Flores-Mendieta, J.J. Torres, M. Neri and A. Martínez).
74. *Precision radiative corrections to the Dalitz plot of baryon semileptonic decays including the spin-momentum correlation of the decaying baryon and the emitted charged lepton*, Phys. Rev. D **72**, 057503 (2005), Erratum-ibid. **72**, 079901 (2005) (with M. Neri, A. Martínez, R. Flores-Mendieta, J.J. Torres).
75. *Mirror matter admixtures in $K_L \to \gamma\gamma$*, Int. J. Mod.Phys. **A21**, 4197-4207 (2006) (with G. Sánchez-Colón).
76. *Precision radiative corrections to the semileptonic Dalitz plot with angular correlation between polarized decaying and emitted baryons: Effects of the four-body region*, Phys.Rev. D **74**, 077501 (2006) (with J.J. Torres, M. Neri and A. Martínez and R Flores-Mendieta).
77. *Precision radiative corrections to the semileptonic Dalitz plot with angular correlation between polarized decaying baryon and emitted charged lepton: Effects of the four-body region*, Phys. Rev. D **75** 097301 (2007) (with M. Neri, A. Martínez, J.J. Torres and R. Flores-Mendieta).
78. *Radiative corrections to the three-body region of the Dalitz plot of baryon semileptonic decays with angular correlation between polarized emitted baryons and charged leptons*, Phys. Rev. D **78** 054018 (2008) (with M. Neri, J.J. Torres, R Flores-Mendieta and A. Martínez).
79. *Radiative corrections to the three-body region of the Dalitz plot of baryon semileptonic decays with angular correlation between polarized emitted baryons and charged leptons: The Initial-baryon rest frame case*, Phys. Rev. D **79** 057502 (2009) (with C. Juárez, A. Martínez, M. Neri, J.J. Torres and R. Flores-Mendieta).
80. *Evidence against manifest right-handed currents in neutron beta decay*, Phys. Rev. D **81** 014030 (2010) (with G. Sánchez-Colón).

Books , chapters in books and reviews

1. *Desintegración beta de bariones: situación actual y perspectivas*, Rev. Mex. Fís. **28**, 481-502 (1982).
2. *The beta decay of hyperons*, Lecture Notes in Physics **222**, 1-173 (1985) (with P. Kielanowski).
3. *High precision corrections to the free neutron beta decay. Topics in Contemporary Physics*, (with J.L. García-Luna) edited by José A. Heras and Rolando V. Jiménez. Instituto Politécnico Nacional 2000, 37-52 (appeared in 2001).

Conference proceedings

1. *Hyperon decays and spectrum generating* $SU(3)$, Proceedings of the international symposium on mathematical physics, January 1976, Vol. 1, 403-420 (with A. Bohm and R. Teese).
2. *Baryon semileptonic decays-Spectrum generating* $SU(3)$ *and the Cabibbo model*, Lecture notes in Physics **94**, 311-316 (1979) Springer Verlag, Heidelberg.
3. *Charmed baryon decays*, Ann, N.Y. Ac. Sc. **520**, 128-131 (1988) (with R. Huerta and P. Kielanowski).
4. *Predictions for charm baryon semileptonic decays in the* $SU(4)$ *limit, the nonrelativistic and MIT bag quark models*, Ann. N.Y. Ac. Sc. **578**, 420 (1989) (with M. Avila, R. Huerta and R. Pérez-Marcial).
5. *Weak Mixings of hadrons*, Proceedings of the 1993 workshop on Particles and Fields. World Scientific, Singapore 1994, 291-297 (with G. Sánchez-Colón and R. Huerta).
6. *A priori mixed hadrons, hyperon non- leptonic decays, and the* $\Delta I = 1/2$ *rule*, AIP Conference Proceedings **359**, 1-6 (Am. Inst. Phys., N.Y., 1996) (with R. Huerta and G. Sánchez-Colón).
7. *Quark mixing possibilities in extensions of the standard model and the electromagnetic current*, AIP Conference Proceedings **400**, 237-243 (Am. Inst. Phys. N.Y. 1997) (with R. Huerta y G. Sánchez-Colón).
8. *Baryon semileptonic decays*, Hyperon Physics Symposium, Hyperon 99, Fermilab (ed. D.A. Jensen and E. Monnier, Sept. 1999).
9. *Strong Interaction corrections to the neutron beta decay and high precision*, AIP Conference Proceedings **531** 386-390 (2000) (ed. A. Ayala, G. Contreras and G. Herrera) (with J.L. García-Luna).
10. *La física de partículas elementales o el alcance de la superconductividad*, Rev. Mex. Fís. **34**, 325-343 (1988).
11. *Correcciones radiativas a la gráfica de Dalitz en decaimientos semileptónicos de hiperons polarizados*, Coloquio de Investigación. Memorias, April 1994, Escuela Superior de Física y Matemáticas, IPN, 97-108 (with R. Flores, A. Martínez and J.J. Torres).

12. *High Energy Physics in Mexico: Historical Sketch and Implications* AIP Conference Proceedings **857**, Part B, 3-10 (2006) (ed. Miguel A. Pérez, Luis F. Urrutia, Luis Villaseñor) (with M.A. Pérez).
13. *Radiative corrections to the Dalitz plot of charged and neutral baryon semileptonic decays with angular correlation between polarized emitted baryons and charged lepton*, AIP Conference Proceedings **1026** 309-311 (2008) (with J.J. Torres, A. Martínez and M. Neri). AIP Conf.Proc.1026:309-311,2008.

Articles for general public

1. *El posgrado y la investigación científica en el Centro de Investigación y de Estudios Avanzados*, Ciencia y Desarrollo 50, 60-70 (1983) (with M.A. Pérez).
2. *El sistema nacional de investigadores*, Editorial in Bol. Soc. Mex. Fís. **1**, 35-35 (1987).
3. *Aprendizaje sobre la marcha: desarrollo del Departamento de Física (1963-1996)*, Avance y Perspectiva **15**, 97-104 (1996).

Internal reports

1. *La física en el Cinvestav*, 1-23 (with M.A. Pérez).
2. *El descubrimiento de la ecuación de Schödringer: audacia, intuición o suerte?*, Revista de Física, 66-68 U. De Sonora, Dec. 1987.
3. *Estudio y propuesta para la preservación de los grupos activos de investigación*, Academia de la Investigación Científica, 1-16 March 1989 (with the rest of the committee).

Augusto García: Some remembrances

R.G. Barrera

Instituto de Física, Universidad Nacional Autónoma de México, Apartado Postal 20-364, 01000 México D.F., Mexico

I am here today because I want to share with you some remembrances about my dear friend and colleague Augusto García. When one asks oneself: who was Augusto García? one has the feeling that the question is not well posed, that it is really hard to find a concrete answer... and that is correct. Nevertheless I think that a possible answer to this question is to say that Augusto García is now the collection of all thoughts and remembrances that all we have about him. This kind of answer brings to my mind the famous Mexican novel Pedro Páramo, when at the beginning the son of Pedro Páramo gets into the town where his father has lived and starts to ask about him, and he gets quite perplexing different views of him depending on whom he asks... an after a while one realizes that the different characters that have met him and tell stories about him... the ones who talk... were remembrances themselves.

I am not in the field of high-energy physics, thus I did not understand very well what was Augusto actually doing in his research... although some times he tried to explain that to me... therefore I prevent you not to expect any technical opinion about the physics Augusto was doing, neither about how clever were his solutions to this or that problem. I am here just to tell you some remembrances about him as a friend and as a fellow student, and in this way you will have another view... another perspective... about him, especially today, when we all got together here to think about him... now that he is gone.

Augusto and I met when both of us were starting to study physics at the National University of Mexico in the Facultad de Ciencias (School of Sciences). It was a four year program in physics and mathematics and after the completion of a thesis one ended up with a professional degree called "Physicist". In that first year all of us were extremely enthusiastic about learning physics and our minds were full of dreams about building up new theories or discovering subtle new physical effects. Augusto was educated in the schools that were founded by the Spanish community who arrived into Mexico at the end of the Spanish civil war, from the defeated side, looking in Mexico for a new place to live. Within this community there were several intellectuals that ended up as university professors, some of them teaching physics and mathematics. This Spanish community was very active and their schools had a quite liberal conception of what education should be, especially for youngsters that were starting a new life in a new country.

It was in 1960 when we started our first year of physics and we were only about eighty students. We were split in two groups taking our lessons in two different classrooms. The School of Sciences had a extremely nice building at the center of the beautiful new university campus, that started to operate just five years prior to our arrival as students. Before that, the university very old buildings were all scattered in the downtown area

CP1259, *First CINVESTAV-UNAM Symposium on High Energy Physics*
edited by J. C. D'Olivo, R. López-Fernández, and M. Á. Pérez

and the new university campus was now the pride of the modernization policies of the government. Right next to the School of Sciences there was the Tower of Sciences, a thirteen story building that embodied the research institutes of astronomy, geophysics, mathematics, physics and chemistry, and right at the top, in the thirteenth floor, there was the office of the Dean of Science with a superb view to the volcanoes. This was the first time that the 20 years old Institute of Physics had a building of its own and this as the place where the offices of our teachers were.

It was in this place with such a stimulating atmosphere where Augusto and I studied physics together for more than four years. Augusto had two very close friends that came with him from the Spanish schools and they were also children from the Spanish refugees. The three of them were very good students, they study together and they were always the first who completed the homework assignments. We "hated" them and I remember that we used to call them the "Real Madrid" team, by recalling the internationally famous Spanish soccer team. At that time I had many, many conversations with Augusto, about physics and about almost anything. We shared many things together especially some fantasies about our future lives as physicists. I remembered that our last exam was quantum mechanics, we studied a lot and finally... after that exam, there was nothing more to do... It was over... we had finished. All the class went together to celebrate this long-waited event and we did it in the most traditional way: drinking beer and eating tacos. Later on I organized a more formal dinner in a classy restaurant, all dressed up, white shirts, ties... and when we were looking for a place to sit... then,, very solemnly... Augusto enters the room with a very beautiful young girl... she was Spanish and her name was Queti... all of us stared at them... feeling... deep down some kind of envy. We learned later that they were engaged and after a few months they got married. Later on she became the mother of his three sons.

The next step was to do the thesis work which ideally should take about a year. In the institute of physics the strongest research topic was nuclear physics with a rather mathematical approach. Augusto was fascinated by particle physics and he chose this area to work a problem with a Spanish professor who was an excellent and bright lecturer who was our instructor in some courses but, as we found out later on we, he was not very active in research. I myself found mathematical physics rather dry and chose to do my thesis on solid-state physics, as it was called at the time. The only problem was that there was only one young researcher at the institute working on that area, and he was leaving to England to do, what we would call now, a postdoctoral stay. So I started to work with a young professor who got his doctorate in Germany, was leaving the field of nuclear physics and wanted to learn some solid state.

Augusto and I started to work very hard on our thesis projects but after several months of trying to make some sense of our work we found out that we were going extremely slowly and the end was not clearly stated. Augusto was planning to go to France to do his doctorate and I had already applied to go to the United States through the Fulbright's scholarship program of the Department of State. But it was absolutely necessary that we should finish on time. That was a must. Then... something happened. Augusto, after a lot of thought, finally decided to change adviser and to finish... or rather start working on a new problem with a young, once more, Spanish professor who was coming back from Princeton. Things started to go better with him, the problem was well posed, he got a lot of interaction with this young professor, he was getting results and he was optimistic

FIGURE 1. Here there are from left to right: Bernardo Wolf, Miguel Angel Jimenez, Teodoro González, Augusto García and Guillermo Aguilar.

about finishing up. There was only one problem. He had not told his former adviser that he was doing his thesis work with somebody else. He was afraid that he could get mad... and he was right... he got really mad... we could say... furious... and when he saw what Augusto has been actually doing he said simply that everything was wrong. Augusto was very, very worried... extremely worried, without knowing what to do. He finally decided to write his thesis with his new results, but it was not possible to get rid of his former adviser out of the Examination Committee, composed by three members of the Faculty. The day before the exam Augusto was extremely nervous, smoking more than usual, and all of us, his friends, trying and trying to raise his spirit, without too much success. And the next day came... It was the usual open oral exam and all of us, his friends, were there to witness a terrifying academic experience. During the exam his former adviser was... how can I say... well... in a couple words I could describe him as: extremely cruel. Even being in the audience was rather uncomfortable... to say the least. Since his former adviser was not the only member of the Committee, Augusto was, after a lot of suffering, finally approved. On my side, I read the book by David Pines, "Elementary Excitations in Solids" which I found fascinating, I wrote a report on this book, as my thesis, and with that I also got, on time, my "Physicist" degree. Here I show now some pictures of Augusto and his fellow classmates when we were studying physics at the School of Sciences at the National University of Mexico, between 1960 and 1965. In Fig. 1, I show a picture which should have been taken around 1962-1963 in front of the

FIGURE 2. This table should have been at Dalmau's reception. On the left side the fourth from front to back is Augusto, the sixth is Rubén G Barrera and the seventh (last) is Dalmau Costa. On the right side, the first from back to front is Teodoro González, the second is Vinicio Serment (one of our instructors), the fourth is Alfonso Mondragón (another of our instructors) and the women and girls who appear here, I do not remember who they were.

Tower of Sciences, and there Augusto appears with some of his classmates. In Fig. 2, I show a picture which should have been taken at the reception given by Augusto's friend Dalmau Costa, when he got his physicist degree. Dalmau's father was well to do and owned two very well-known, fancy restaurants in Mexico City, at the reception what I most clearly remember was, that the food was... excellent. It should have been around 1964-1965. In Fig. 3, I show a picture where in the back row, standing, there is Augusto and some of his fellow classmates, and in the front row, the persons who are sitting are at the time, high-rank university officials: the Dean of Science (Alberto Barajas), the President of the University (Nabor Carrillo) and the Director of the Institute of Physics (Carlos Graef). This picture should have been taken at Dalmau's reception.

The next step was to go for a PhD. Augusto left for France and I left for the US, to the University of Illinois at Champaign-Urbana because I wanted to do work with David Pines. At that time Illinois was regarded as one of the best solid-state departments in the US. During a year Augusto and I exchanged some postcards and a couple letters and he told me that he had to prepare an exam called d'étude à profondie, which sound very impressive. We met again, a year later, at my parent's place, during Christmas Holidays. Augusto had already a son, Enrique, and he was not enthusiastic at all of

FIGURE 3. In the back row, standing, from left to right: Augusto García, Rubén G. Barrera, Dalmau Costa and Teodoro González. In the front row, sitting, from left to right: Alberto Barajas, Nabor Carrillo and Carlos Graef.

going back to France. Somehow he did not like it. I told him about my very positive experience at Illinois and encourage him to move to the US. I learned later that he was going to start his PhD at University of Chicago. Well, after all, Champaign-Urbana and Chicago were not very far from each other… the train cost about six dollars and the ride was about three hours. In those years I was still single and Augusto had another son, who was called Augusto, a name that I thought was not very imaginative. In those years I visited him several times. Augusto, Queti and his two sons lived in an apartment that the university provided to married graduate students. It was in a big building and I remember I had to take the train to the station in the 53th street and then walked for a few blocks. During the winter time the weather was… miserable… too much snow, too much wind, and extremely cold. When I arrived at Augusto's place I used to see diapers hanging all around the living room, very active children running and cradling around and a very tired mother trying to have some rest. But Augusto was very much satisfied of being in the Physics Department of a quite prestigious university. He had a good relation with his adviser and his fellow students. In one of my visits to his place he took me apart and told me in low voice… Rubén, I am very worried… why? I asked him, and he answered… also in low voice… because Queti is pregnant and the only income I have is my scholarship… and I had to agree that it was good reason to be worried. Nevertheless it turned out to be a false alarm, and with two children he was capable to manage his financial situation up to the completion of his PhD degree. I remember also that once the whole García family visited me at Urbana. I was very much impressed and somehow touched, when I saw them coming down from the train with quite a number of bags, toys... and... a baby stroller. We had a very nice time... after all... things were working well.

Several years had to pass before the completion of our PhD degrees. After those years as students in the US, we both... eventually... came back to Mexico. At that moment our lives start to diverge but keeping always the warmth that was built up during the time we were students together. Augusto had a third son, Pablo, and he got a position at a prestigious research institution (Cinvestav) in the northern part of Mexico City. From my side, I got a position at the Institute of Physics of the National University of Mexico, in the southern part of the city. The distance between the two research institutions was about 30 kilometers, but the probability of getting into a traffic jam was very close to one, so it would take more than an hour to go from one place to the other, almost independently of the mean of transportation. Thus we used to meet not very often, but when we did we used to have interesting conversations not so much about physics, but mainly, besides personal issues, about the government policies directed towards the development of science in Mexico. Also, the path that has to be followed to built a strong graduate school in physics as well as the education of young researchers, were issues of his main concern. I will add here, that Augusto built up a nice family, his first son, Enrique became a prestigious ophthalmologist, who works now in the US, and his third son Pablo got a doctorate degree in England, in musical composition, and he is now working in Mexico.

It might now seem that my story is getting to an end. But life has many surprises... and there are facts that we do not understand and we usually classify them under a generic term called: luck. One morning I was in my office and I got a telephone call from a young researcher asking me to go to his lab, because he had some interesting results he wanted me to see, concerning the index of refraction of milk. Since I have been working in the field of optical properties for many years this was not a surprise, but that the voice on the other side of the line came from Augusto García... Junior... that was quite a surprise. Augusto Junior had studied physics in Mexico and got a PhD in EE in the US, following the steps of his father. He was working in optics and had a position at the National University of Mexico, my university, at a research center dedicated to technological innovation. After our first encounter I realized that Augusto Junior was an extremely competent experimentalist with a strong theoretical knowledge that allowed him to do sophisticated calculations. We all know that it is not easy to collaborate with somebody, sometimes scientific collaboration can be as or even more complicated than marriage. Well in the case of Augusto Junior and I we have had, in the last years, an extremely fruitful collaboration that has lead to the creation of a research group in the area of optical properties of colloids. It is quite curious to collaborate with somebody that I met in Augusto's place in Chicago when this somebody was wearing diapers. But that is the way life is all about. Thus I will finish saying that for me, from my very personal perspective, one of the main legacies of Augusto García was... Augusto García. Thank you very much Augusto and... rest in peace.

EXPERIMENTS AND ACCELERATORS

Sensitivity of a Single Water Cerenkov Detector to Monitor Forbush Decreases

A. Bahena* and L. Villasenor†

*Facultad de Ciencias Fisico-Matematicas, UMSNH, Morelia, Mexico
†Institute of Physics and Mathematicas, UMSNH, Morelia, Mexico

Abstract. We describe a simple experimental setup to measure the rate of arrival of muons at the surface of the Earth by using a single water Cerenkov detector and home-made electronics. We find a strong anti-correlation between the muon rates averaged over one-hour periods and the atmospheric pressure, with a measured correlation coefficient of -0.67% per hPa. After applying this correction we achieve sufficient sensitivity to observe long term (hours)variations in the averaged muon rates which are greater than 2%. Forbush decreases as big as 4% have been observed with muon detectors located at similar magnetic rigidities compared to Morelia, therefore our experimental setup will detect Forbush decreases as soon as the Sun enters into a more active phase.

Keywords: Water Cerenkov detector; Cosmic ray muons; Forbush decrease
PACS: 96.50.S-; 41.60.Bq; 96.60.qe

INTRODUCTION

The Sun possesses an eleven-year period of activity, at its peak it may eject intense clouds of magnetized plasma; the biggest of these solar coronal mass ejections may in turn travel toward the Earth sweeping some of the galactic cosmic rays away from Earth. This in turn gives rise to a decrease in the flux of cosmic rays that arrive at the Earth. This effect is called Forbush decrease [1] after the physicist Scott E. Forbush who discovered it in 1937. Forbush decreases of secondary cosmic rays are detected by worldwide networks of monitors of neutron fluxes and, more recently, detectors of fluxes of secondary muons. The Forbush decrease is usually observable on Earth within a few days after the coronal mass ejection by the Sun; the decrease in the flux of secondary cosmic rays takes place over the course of a few hours. A practical application of the early detection of Forbush decreases is to use them as precursors of the perturbations of the magnetosphere of the Earth caused by the direct interaction of the passage of the solar magnetic storms.

Neutron monitors are the preferred instruments to monitor Forbush decreases because they respond to lower energy cosmic rays that suffer more heliospheric modulation compared to muon detectors. However, the advantage of muon monitors is that they are cheaper and easier to operate. Since they respond to the same phenomenon, i.e., the detection of secondary cosmic rays on the ground, the data from muon counters can be readily compared with the data from neutron monitors available in real time on the World Wide Web [2]. In this paper we study the sensitivity that can be achieved with a single water Cerenkov detector of small dimensions to monitor Forbush decreases by measuring the flux of secondary muons.

CP1259, *First CINVESTAV-UNAM Symposium on High Energy Physics*
edited by J. C. D'Olivo, R. López-Fernández, and M. Á. Pérez

FIGURE 1. Photograph of the water Cerenkov detector used for this experiment (out of the two detectors shown we used the one on the left side). It consists of a polyethylene cylinder of 1.1 m diameter filled with purified water up to a height of 1.1 m and an 8" photomultiplier located at the center slightly immersed and looking downwards into the water volume.

In the next section we describe the experimental setup and the home-made electronics we used. Next we present the data analysis and discuss the results we have obtained recently. Finally, we conclude that by using a relatively small detector of 1 m of height and 1.1 m of diameter we reach a sensitivity of the order of 2% to variations on the rate of arrival of secondary muons provided we average the rates over periods of 1 h and correct the data to take into account the variations on the muon rates due to changes in atmospheric pressure.

DESCRIPTION OF THE EXPERIMENTAL SETUP

The water Cerenkov detector we use consists of a polyethylene cylinder of 1.1 m diameter filled with purified water up to a height of 1.1 m. The inner surface of the tank was covered with a sheet of a highly reflective material called tyvek, which has a diffuse reflectivity of about 90% in the ultraviolet region of the EM spectrum.

A single photomultiplier of 8" of diameter, EMI 9353KB, was located at the center of the tank with the photomultiplier slightly immersed in the water and looking downwards into the water volume. We covered the detector with additional sheets of polyethylene to ensure that it was completely free of light leaks, see Figure 1. The photomultiplier signal was transmitted via RG58 coaxial cable to a digital oscilloscope, Tektronix TDS220. The typical amplitude of the photomultiplier signal for muons crossing the detector was

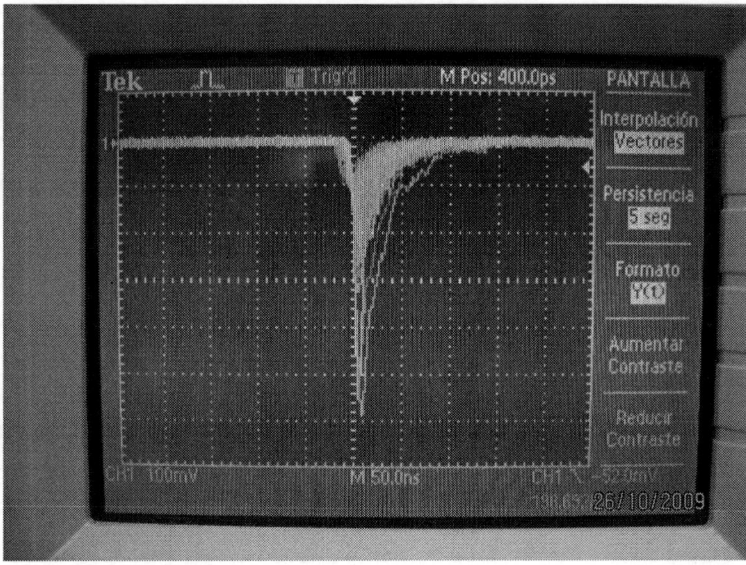

FIGURE 2. Typical amplitudes of the photomultiplier signal for muons crossing the detector. The vertical scale is 100 mV/div and the horizontal scale is 50 ns/div.

about 200 mV, see Figure 2, this amplitude corresponds to an energy deposition of 200 MeV for our detector of 1 m height (the energy loss of minimum ionizing muons is 2 MeV/cm).

Figure 3 shows a picture of the home-made electronics [3] that we used to process and acquire the signal from the photomultiplier into our PC. It consists of a commercial FPGA board (Digilent D2FT) with a 400K-gate Xilinx Spartan 2E XC2S400E FPGA in an FT256 package that provides 168 user I/Os; a custom-made ADC daughterboard that digitizes the signal from the photomultiplier with a sampling period of 10 ns; this ADC uses the AD9214 10-Bit sampling analog-to-digital converter [4]. Since it is crucial to measure the atmospheric pressure with high precision, we incorporated a pressure/temperature sensor into our electronics (HP03) with a resolution of 15 Bits and accuracy of 0.5 HPa [5]. We also included a GPS receiver in our acquisition electronics (Motorola Oncore UT+) to measure time intervals and UTC time with high precision [6]. Figure 3 shows the FPGA board with the ADC daughterboard connected (upper left side), the pressure sensor connected on the lower right side and the GPS receiver shown in the middle right side of the picture. Since the FPGA board does not include an RS232 port, we used a serial converter interface (Digilent PModRS232) [7] shown in the lower right side of Figure 3.

We wrote the firmware of our electronics in VHSIC hardware description language (VHDL), where VHSIC stands for very-high-speed integrated circuit. The tasks that our data acquisition system performs are the following:

- Get the pulse-per-second (PPS) signal from the Oncore UT+ GPS receiver.

FIGURE 3. Photograph of the electronics used to measure the rates of arrival of secondary cosmic ray muons, atmospheric pressure, temperature and GPS time. This acquisition system communicates with a PC by using an RS232 port.

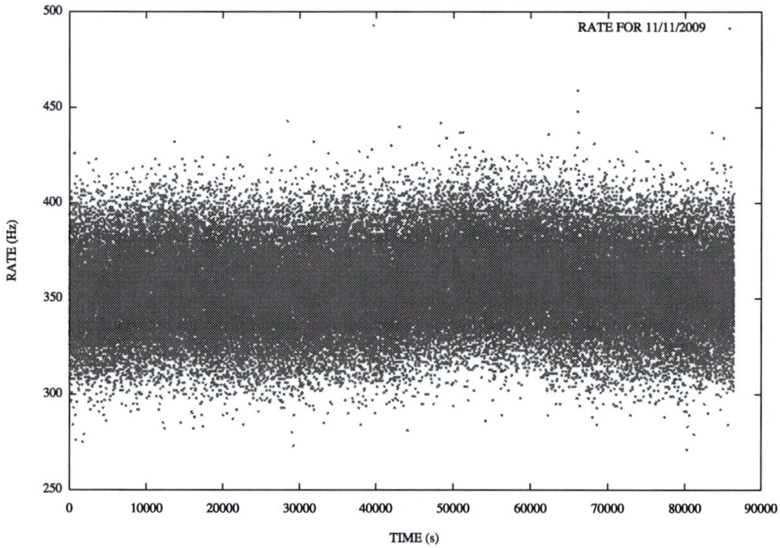

FIGURE 4. Plot of the rate measured every second during a whole day (November 11, 2009). The vertical axis is the muon rate in Hz for a threshold of 22 mV. This threshold is equivalent to about 22 MeV of energy deposition of the muon. The horizontal axis is the local time in seconds.

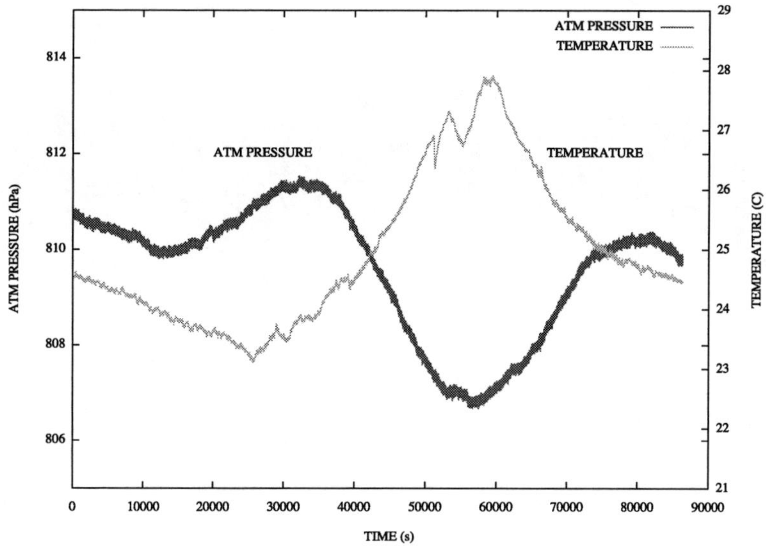

FIGURE 5. Plot of the atmospheric pressure (left axis) and the temperature (right axis) measured every second during a whole day (November 11, 2009). The horizontal axis is the local time in seconds.

- Measure the rates in one-second intervals from the ADC board at a threshold of 22 mV which roughly corresponds to a muon deposition energy of 22 MeV or 0.11 VEM (Vertical Equivalent Muon)
- Get the GPS time from the Oncore UT+ GPS receiver
- Read the atmospheric pressure from the HP03D sensor in synchrony with the PPS signal
- Send all the above data in binary format to the PC in synchrony with the PPS signal by using the RS232 port

On the PC side we use Linux tools to perform the following tasks:

- Capture the data on the RS232 port
- Automatically create binary files every day
- Convert the binary files to decimal format
- Analyze and plot the data.

RESULTS AND DISCUSSION

Figure 4 shows the rate measured every second during a whole day (November 11, 2009), this was a typical day during a period of low solar activity. We have checked that the distribution of these muon rates is given by a Gaussian distribution with a standard deviation equal to the square root of the mean value; this means that the muon detection

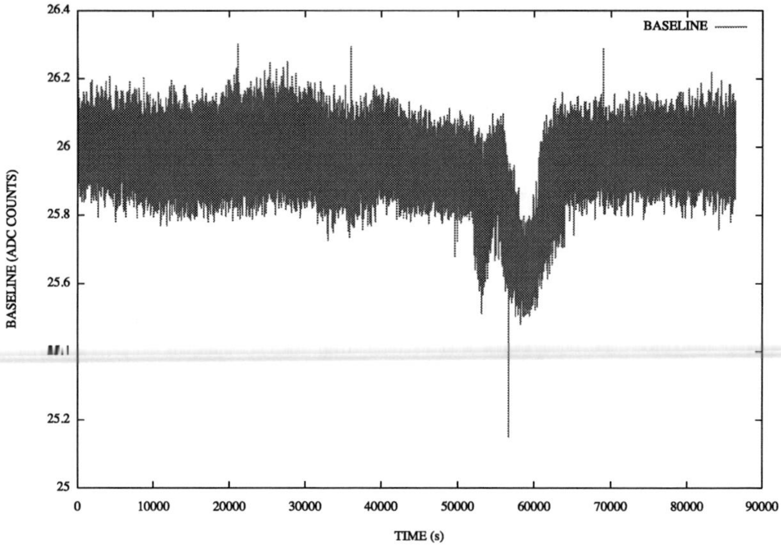

FIGURE 6. Baseline of the photomultiplier signal measured at the beginning of every second. The units of the vertical axis are ADC counts. An ADC count is equal to 2 mV.

is a random process with a constant probability per unit time, i.e., a Poisson process. In turn this constitutes an important test of the stability of the detector.

Besides measuring the rate every second, our custom-made electronics and acquisition system also records the atmospheric pressure and the temperature every second. Figure 5 shows the atmospheric pressure (left axis) and the temperature (right axis) for the same day (November 11, 2009). The atmospheric pressure exhibits two characteristic peaks per day while the temperature only one. In addition, we also record the baseline every second, and, more importantly, our discrimination threshold is anchored to the measured baseline. Figure 6 shows the baseline measured every second for the same day (November 11, 2009). The displayed value is the average of the baseline measured repeatedly for about 10 ms at the beginning of every second. The units of the vertical axis in Figure 6 are ADC counts, with an ADC count equal to 2 mV.

Figure 7 shows the strong anti-correlation between the average of the muon rates over periods of one hour (left axis) and the atmospheric pressure (right axis). Note that the right axis in Figure 7 has been inverted to make this anti-correlation more evident. The period displayed in this Figure corresponds to 13.3 days of data taking in early November, 2009. The statistical error on the average is equal to the standard deviation of the rate measured every second, roughly equal to the square root of the mean value measured every second, i.e., about 19 Hz, divided by the square root of 3600, i.e., the number of seconds in one hour. Therefore, the statistical error on the rate displayed in Figure 7 is negligibly small, about 20 Hz divided by $\sqrt{3600}$, i.e., 0.3 Hz.

Figure 8 shows the scatter plot of the hourly average of the rate (vertical axis) versus the hourly average of the atmospheric pressure (horizontal axis). The anti-correlation

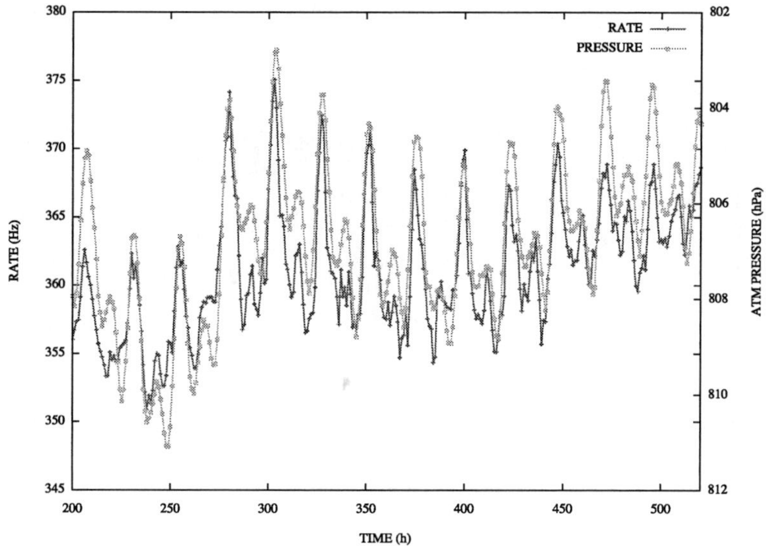

FIGURE 7. Anti-correlation between the average of the muon rate over periods of one hour (left axis) and the average of the atmospheric pressure (right axis). Note that the right axis has been inverted to make this anti-correlation easily evident.

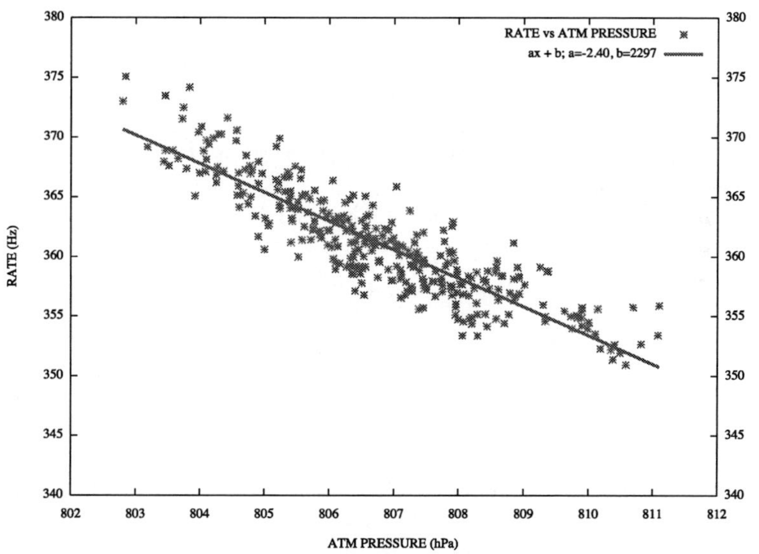

FIGURE 8. Scatter plot of the hourly average of the muon rate (vertical axis) versus the hourly average of the atmospheric pressure (horizontal axis). The solid line is a least-squares fit, with a slope of -2.40 ± 0.07 Hz/hPa. The measured correlation coefficient is therefore equal to 0.67% per hPa.

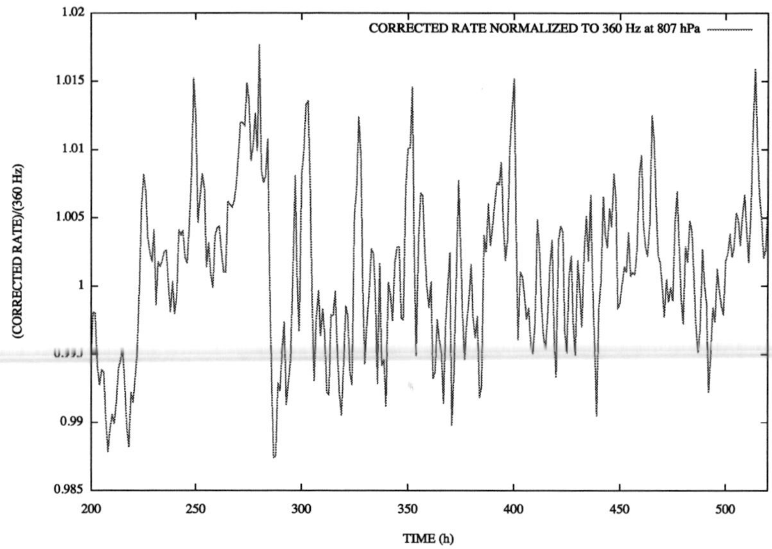

FIGURE 9. Muon rate averaged over one-hour periods corrected to take into account the anti-correlation with atmospheric pressure. The horizontal axis is in local hours for data taken in November, 2009. A total of 13.3 days are displayed from November 8 to November 21, 2009.

between theses averages is clearly evident. The solid line is a least squares fit, with a slope of -2.40 ± 0.07 Hz/hPa. This anti-correlations is the well-known barometric effect on the flux of secondary cosmic rays; it is due to the larger or smaller mass of air in the path of the secondary cosmic rays when the atmospheric pressure is high or low, respectively, giving rise to a higher or smaller attenuation of the secondary muons, i.e., a lower or higher flux at ground level.

Figure 9 shows the rate averaged over one-hour periods and also corrected to take into account the changes in atmospheric pressure. The rate values correspond to average rates projected to a fixed atmospheric pressure of 807 hPa using the linear anti-correlation shown in Figure 8. This correction simply uses the measured slope between averaged rate and atmospheric pressure, i.e., -2.40 ± 0.07 Hz/hPa. Note that the variations on the corrected average rates are of the order of $\pm 1.5\%$, whereas the variations on the uncorrected averaged rates are of the order of $\pm 4.2\%$, i.e., we gain a factor of about 2.7 in the sensitivity of our single water Cerenkov detector to Forbush decreases by applying this correction.

The daily variations in the corrected rate, see Figure 9 are in part due to the temperature variations shown in Figure 5 and in part due to the rotation of the Earth which allows the detector to successively view regions of greater or lesser cosmic ray activity.

CONCLUSIONS

A single water Cerenkov detector of 1.1 m of diameter filled with purified water up a height of 1 m has been shown to possess sufficient sensitivity to observe variations greater than 2% in the muon rate averaged over periods of one hour. Forbush decreases as big as 4% have been observed with muon detectors located at similar magnetic rigidity compared to Morelia [8]. Our instrument is continuously monitoring the Sun activity since October, 2009 with stable operation. The Sun has been in a low activity phase during this time, but it may enter into a more active phase at any time. When this happens we will undoubtedly detect Forbush decreases with the instrument described.

ACKNOWLEDGMENTS

The authors thank CONACyT and the Coordinacion de la Investigacion Cientifica from UMSNH for supporting this project.

REFERENCES

1. http://astronomy.swin.edu.au/cms/astro/cosmos/F/Forbush+Decrease
2. See for instance the on-line data from the neutron monitor of UNAM, Mexico, at http://www.geofisica.unam.mx/infra/observ.html
3. Gonzalo Perez, PhD Thesis, FCFM-BUAP, Puebla, Mexico, 2005
4. http://www.analog.com/en/analog-to-digital-converters/ad-converters/ad9214/products/product.html
5. http://www.futurlec.com/HP03D.shtml
6. http://www.eecis.udel.edu/ mills/ntp/html/drivers/driver30.html
7. http://www.digilentinc.com
8. Pierre Auger Collaboration, Proceedings of the 31st ICRC, Lodz 2009, arXiv:0906.2347, p. 28, 2009.

Searching for Dark Matter with a Bubble Chamber

Peter S. Cooper

Fermi National Accelerator Laboratory
PO Box 500 Batavia, IL 60510, USA

Abstract. The Chicagoland Observatory for Underground Particle Physics (COUPP) recently reported first Physics results from an engineering prototype dark matter detector consisting of a 1 liter, room temperature, heavy liquid bubble chamber. These results close the last open window in low mass spin dependent WIMP scattering for a conventional WIMP interpretation of the DAMA annual modulation signal. I will discuss these results and the detector techniques which enabled them.

Keywords: Bubble Chamber, Dark Matter
PACS: 29.40.-n, 95.35.+d

AUGUSTO GARCIA

This symposium is held to honor the memory and career of our colleague Augusto Garcia. Augusto and I knew each other from conferences, however, we mostly worked one step removed from each other. I spent much of the 1980's doing experimental work in hyperon beta decays where Augusto was one of the recognized world theoretical experts. I learned that subject in no small measure from Augusto and Piotr Kielanowski's publications. In the late 1980's Joe Lach of Fermilab and I were local advisers for Augusto's grad student Antonio Morelos, now professor of Physics at San Luis Potosi, who came to Fermilab under a graduate student exchange program of that time.

Only once did we get to really sit down and do physics together. Augusto and Rodrigo Huerta came up to Fermilab around 2000 to discuss a new, and quite unusual, physics idea they had. They enlisted Bill Bardeen and I and we retired for the afternoon to a small conference room to talk through their ideas. It was - shall we say - intense, and I hope helpful to them. It was one of the few days I can distinctly remember as having actually done some physics in contrast to how we usual spent almost all of our time. Surely theorists get to do more of that but for an experimentalist like me is was a singular joy.

INTRODUCTION

The direct observation and characterization of cold dark matter; 85% of the matter in the universe, is one of the most pressing quests in physics today. After more than 30 years as a working particle physicist studying the properties of matter I consider it an embarrassment to learn that for every gram we've studied there are 5 grams about which all we know is that it exists and gravitates.

CP1259, *First CINVESTAV-UNAM Symposium on High Energy Physics*
edited by J. C. D'Olivo, R. López-Fernández, and M. Á. Pérez

An experiment to directly detect and begin to characterize the dark matter is a formidable undertaking. The Weakly Interacting Massive Particle (WIMP) hypothesis is a most attractive candidate for the dark matter: a lightest super-symmetric neutral particle in the $10-1000GeV/c$ mass range is fashionable in the LHC era. The search techniques for WIMPs all involve observing nuclear recoils from the collision of a WIMP with the nucleus in a detector. The WIMP velocity is dominated by velocity of the sun around the center of our galaxy; typically $\beta = 0.001$ producing nuclear recoils with kinetic energies of 10's of KeV.

Any detector has to be highly sensitive to KeV nuclear recoils and have excellent rejection and/or suppression for all other background processes. Cross section limits much below picobarns must be achieved. Any successful detector must be scalable and cheap. The most pressing backgrounds are from radioactive decay so high radio-purity materials are also required.

I'll not believe the first claim of the direct observation of dark matter even if my collaboration is the one which makes it! The detection signatures are all sufficiently tenuous, and the possibly for un-understood backgrounds much too high, for any single observation to be conclusive. A critically important requirement is to observe the interaction of dark matter with different nuclei producing a clear pattern of interaction cross-sections in order to conclusively distinguish a true signal from a mysterious background.

Our detection technique of using a bubble chamber sensitive to only nuclear recoils has this advantage. You can make any liquid which will boil into a bubble chamber working fluid. Changing the nuclear makeup of the working fluid is straight-forward and cheap.

Without knowing the properties of WIMPs the WIMP-nucleus interaction can only be describe in general terms. There is a spin independent (SI) interaction for all WIMPs and nuclei and a spin dependent (SD) interaction possible if both the nucleus and WIMP have non-zero spin.

The spin independent (SI) interaction, given by equation[1], is described in terms of proton and neutron coupling constants ($f_{p,n}$). Notice that this is a coherent interaction with the whole nucleus at these very low momentum transfers (q^2) modified by a nuclear form factor, $F(q)$.

$$\frac{d\sigma_{\chi^N}^{SI}}{dq^2} = \frac{1}{\pi V^2}[Zf_p + (A-Z)f_n]^2 F^2(q) \tag{1}$$

The spin dependent interaction, given by equation[2], is described in terms of proton and neutron coupling constants ($a_{p,n}$). This interaction is between a WIMP and a single nucleon spin. There is no coherence since the final nuclear state is distinguishable.

$$\frac{d\sigma_{\chi^N}^{SD}}{dq^2} = \frac{8G_F^2}{(2J+1)V^2}[a_0^2 S_{00}(q) + a_0 a_1 S_{01}(q) + a_1^2 S_{qq}(q)] \tag{2}$$

$$a_p = \tfrac{1}{2}(a_0 + a_1) \quad a_n = \tfrac{1}{2}(a_0 - a_1)$$

BUBBLE CHAMBERS AS DARK MATTER DETECTORS

A bubble chamber is an excellent detector of nuclear recoils. Furthermore by reducing the superheat (e.g. lowering the temperature to just above the boiling point) only nuclear recoils are capable of nucleating a bubble, making the chamber insensitive to other forms of radiation. The trick is to eliminate trapped gas and surface roughness which nucleate the bubbles in, for example, a glass of champagne. After this is accomplished a bubble chamber can be expanded indefinitely in the absence of sources of nuclear recoils.

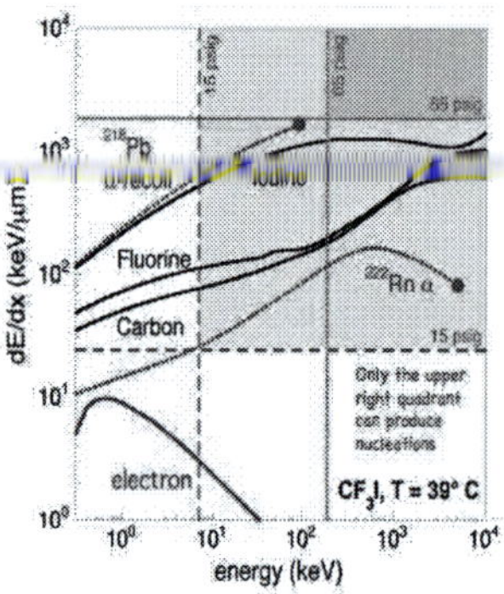

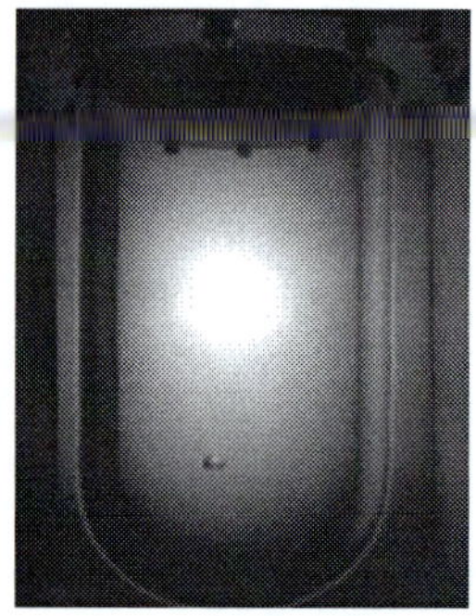

FIGURE 1. Bubble chamber threshold for bubble nucleation as a function of dE/dx and E for various radiations [left]. A typical single bubble "WIMP" like event [right].

A superheated fluid is in a metastable state. If a very small volume of fluid is boiled by some energy deposition the free energy of this proto-bubble as a function of the bubble radius has two terms, a negative on due to the latent heat of evaporating a volume of fluid and a positive one for the work done against the surface tension to increase the surface area. This leads to a critical radius below which bubbles collapse and above which they grow without bound until the whole volume of fluid boils to a gas. This is the Seitz hot spike theory of bubble chamber operation in words.

The threshold for bubble formation requires a sufficient amount of energy in a sufficiently small volume in order to form a bubble. At low superheat, where we operate, a minimum ionizing particle deposits more than enough energy but never in a small enough ($\sim 100nm$) volume to nucleate a bubble. Nuclear recoils with low energy depositions but very large dE/dx always do so. The response of the COUPP bubble chamber with CF_3I as a working fluid is shown in figure(1). The threshold can be trivially adjusted by varying the expanded pressure at constant temperature to adjust the nuclear recoil energy threshold anywhere above $5KeV$. Note that we are fully efficient for C, F and I nuclear recoils while electrons (β's) are completely inefficient. The efficiency for α's turns off in the middle of our operating range. A typical single bubble event is also shown from an α decay in the liquid producing both a $\sim 5MeV\,\alpha$ and an $\sim 80KeV$ nuclear recoil, either of which will nucleate a bubble.

The COUPP bubble chamber is located in the Fermilab MINOS near detector hall at a depth of $300m$ water equivalent. The volume of CF_3I working fluid is 1 liter with a mass of $2Kg$. The CF_3I is contained in a quartz vessel attached to a stainless steel bellows. The balance of the interior volume is filled with ultra-pure water to form a sealed pressurized volume. This assembly is suspended in a pressure vessel filled with a fluid which is compressed and decompressed using conventional hydraulic techniques. The working fluid is viewed with cameras through two orthogonal ports. The chamber is expanded to a working pressure in the $10 - 60psig$ range and a 50Hz stream of video images is compared until a significant difference is seen (e.g. a bubble). The chamber is then compressed to $180psig$ for 30 seconds to collapse the bubble and stabilize the chamber. The previous pictures from the cameras and chamber state information are recorded as an event. Typical chamber live-times are $\sim 85\%$.

The data are analyzed by finding the bubbles and categorizing events into bulk single bubbles, wall or surface events, and multi-bubble events. The livetime as a function of pressure is computed. Results are presented as counts per day in a pressure bin.

BACKGROUNDS AND CALIBRATION

Separate data sets were taken with external photon and neutron sources to measure efficiency background rejection.

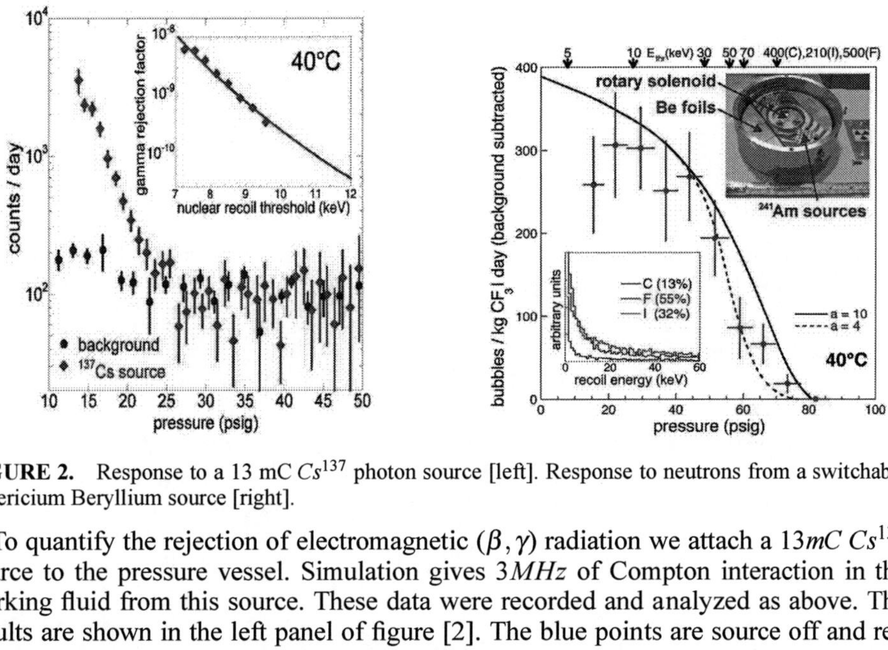

FIGURE 2. Response to a 13 mC Cs^{137} photon source [left]. Response to neutrons from a switchable Americium Beryllium source [right].

To quantify the rejection of electromagnetic (β, γ) radiation we attach a $13mC\ Cs^{137}$ source to the pressure vessel. Simulation gives $3MHz$ of Compton interaction in the working fluid from this source. These data were recorded and analyzed as above. The results are shown in the left panel of figure [2]. The blue points are source off and red source on. The inset translates to a rejection factor as a function of equivalent nuclear recoil threshold. These rejections are in the $10^8 - 10^{10}$ range for $7 - 10KeV$ thresholds.

The right panel of figure [2] shows a similar study using a switchable Americium Beryllium neutron source which produces about $50Hz$ of neutrons. The source is shown in the inset. The response compared to an absolute, ab-initio simulation demonstrates good agreement and full efficiency. The lower inset shows that these neutron induced nuclear recoil events are dominated by the lowest energy recoils.

The largest class of events are bubbles on the quartz walls. The spacial distribution of these bubbles clearly map out the walls of the vessel. These events turn off in pressure or threshold just as predicted for α particles. They are consistent with α's from radioactive decays in the quartz near the inner surface. They are isolated geometrically with a fiducal volume cut and cause increased dead time but are not a background.

RESULTS

The bulk bubble rate is completely consistent with the decay of Rn^{222} and its two subsequent α decay daughters. All three of these α decays have recoil energies near $80KeV$. The time difference between bulk bubble events is shown in the left plot of figure[3]. The $3.1min$ half life of the Po^{218} Radon daughter is clear. As the two lower plots show the bulk bubble rate goes to zero above $80KeV$ as expected for a line source in a threshold detector.

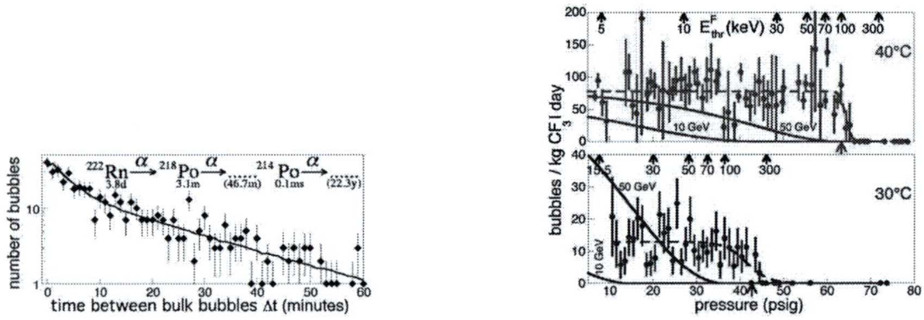

FIGURE 3. Time difference between adjacent bulk bubble events [left]. Bulk bubble rate per Kg per day 40C with 3pb WIMP templates [right upper]. Bulk bubble rate per Kg per day 30C with 3pb WIMP templates [right bottom].

In order to generate limits in the presence of a known background we describe the background from Radon decays with sigmoid function of pressure and fit for the null hypothesis (no signal) to determine the parameters describing the threshold turn-off at $\sim 80KeV$. We then produce template integral WIMP spectra making conventional assumptions for the galactic halo model. Examples for SD couplings with a $3pb$ cross-section for two WIMP masses are shown in figure[3] for both $40C$ and $30C$ data. We then fit the data of figure[3] to the sum of each template and a constant background sigmoid. $90\%CL$ level limits for the amount of WIMP signal allowed are derived directly from these fits.

Spin dependent 90% CL limits are shown in figure[4] for proton [left] and neutron [right] couplings as a function of WIMP mass. These results are dominated by SD couplings to the the three Fluorines in our CF_3I. The left plot in figure[5] shows the limits in proton and neutron coupling constants for a $50GeV/c^2$ WIMP. These results close the last open window for explaining the DAMA results in terms of these conventional models. These results have been published in *Science*[1] where further details may be found.

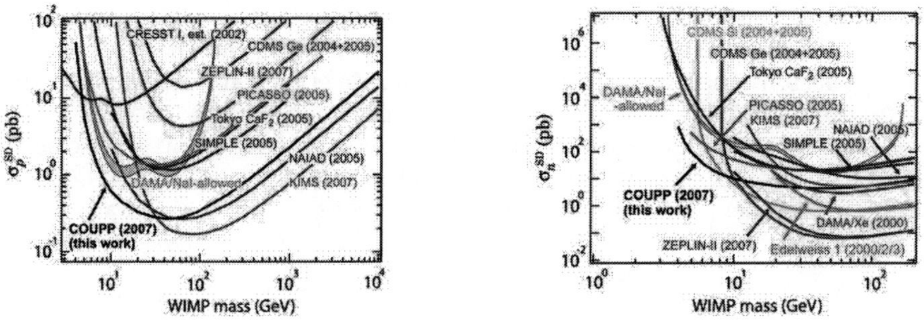

FIGURE 4. Spin dependent cross-section exclusion limits (90% CL) for WIMP-proton [left] and WIMP-neutron[right] scattering.

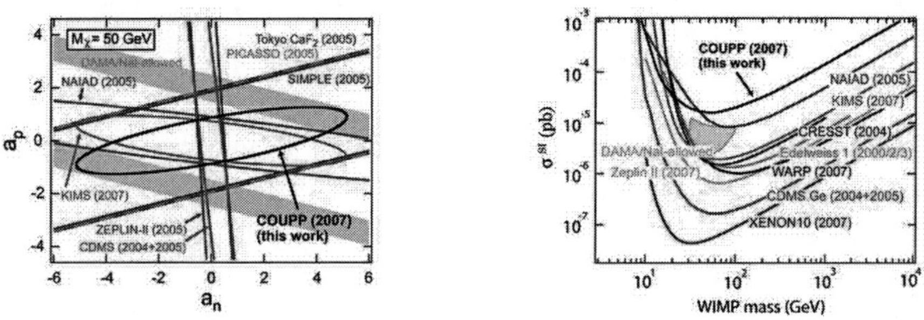

FIGURE 5. Proton vs neutron spin dependent WIMP coupling exclusion limits (90% CL) for a $50GeV/c$ WIMP [left]. Spin independent WIMP-nucleus cross-section exclusion limits (90% CL) [right].

Spin independent 90% CL limits are shown in the right figure[5] as a function of WIMP mass. These are not yet competitive with other experiments. We need to get rid of the Radon background before we can compete with CDMS, *et.al.*

FUTURE PLANS

The 1 liter chamber was reconstructed and refilled with a metal seal in place of the O-ring which was the source of Radon. That background is now gone. We have rebuilt the chamber again with a 2 liter synthetic silica jar with much lower contamination of U and Th decay chain daughters and the high rate of wall events are now gone too. We have also built a good muon system with the 2 liter chamber in the underground MINOS hall at Fermilab to tag and characterize events from muon induced neutrons. Improved limits from the 2 liter chamber will be announced while this report is in press.

We presently also have a completely new 10 liter chamber, constructed at the KICP and a new 30 liter chamber designed built at Fermilab; each with synthetic quartz jars. The former is operation in the Chicago sewer tunnels taking data now and the latter will be operated in the underground MINOS hall at Fermilab beginning in the Spring of 2010 to demonstrate and control the next level of backgrounds.

We have proposed to move the 30 liter chamber to SNOLAB and will do so soon after we have demonstrated successful low background operation at Fermilab. The potential physics reach of this detector in the SNOLAB deep underground site to 2-4 orders of magnitude below the limits shown here. When successfully operated deep underground the 30 liter chamber should be the most sensitive Dark Matter detector in the world for both spin-dependent and spin-independent interactions.

ACKNOWLEDGMENTS

This work is supported by the National Science Foundation through grants to the KICP at the University of Chicago and the Indiana University South Bend, and by The US Department of Energy through Fermilab.

REFERENCES

1. E. Behnke, *et. al.*, *Science* **319**, 933, 2008.

Mexican contribution to ALICE and first data analysis[1]

Eleazar Cuautle[2]

Instituto de Ciencias Nucleares, Universidad Nacional Autónoma de México, Apartado Postal 70-543, México Distrito Federal 04510, México.

Abstract. On November 2009 the Large Hadron Collider produced the first $p+p$ collisions. These, together with the upcoming $Pb+Pb$ data, open up the possibility to answer some of the intriguing questions regarding the Standard Model and likely bring to light new phenomena. In this work I present a short review of the mexican participation in the ALICE experiment, focusing on the detector building and phenemological as well as Monte Carlo simulation work regarding the proton as well as the heavy ion program. I also briefly mention some of the wide spectrum of possible first topics that can be analyzed.

Keywords: heavy ion collision, jets, energy loss
PACS: 13.83.-a, 13.83.Ce, 13.87.Fh, 25.75.Bh

INTRODUCTION

ALICE (A Large Ion Collider Experiment) [1] was conceived as a general-purpose experiment, aimed to study the strong interaction sector of the Standard Model at the extreme values of energy density and temperature reached in heavy ion collisions. As such, it is capable of measuring simultaneously properties of the bulk (soft hadronic, large cross-section, physics) and of rare probes (hard, small cross section, physics). In particular, ALICE is able to track and identify particles from very low (100 MeV/c) up to fairly high (100 GeV/c) transverse momenta in an environment of extreme particle density. Theoretically founded predictions for the multiplicity in central $Pb+Pb$ collisions at the LHC range at present from 2000 to 6000 charged particles per unit of rapidity, at mid-rapidity [2, 3] . The ALICE detectors are designed to cope with multiplicities up to 8000 charged particles per unit of rapidity, a value which ensures a comfortable safety margin. The ALICE detector is built with dimensions 16x16x26 m^3 and total weight of approximately 10000 Tons. ALICE consists of a central barrel part, which covers polar angles from 45^0 to 135^0. Among several subdetectors, we can mention the muon arm,

---

[1] This contribution is dedicated to the memory of Augusto García who, in spite of being a theoretician, always encouraged youngs physicists to consider becoming experimentalists. In fact, the first experimental HEP groups in Mexico were promoted by Augusto among a handful of other colleagues, who helped them to start their training in FERMILAB, DESY and CERN. After many efforts, we can say that there are already established mexican groups in several of the major experiments in the world, one of the largest ones being the ALICE's one. This group has wide interests that include simulation and data analysis as well as design and construction of detector elements, which Augusto was very proud of.

[2] On behalf of the ALICE Collaboration

CP1259, *First CINVESTAV-UNAM Symposium on High Energy Physics*
edited by J. C. D'Olivo, R. López-Fernández, and M. Á. Pérez

the FMD, T0, V0, ACORDE and the VHMPID. In the next section I describe the last 3 of the former.

MEXICAN CONTRIBUTION TO THE DETECTORS OF THE ALICE EXPERIMENT

The V0 detector [4] is a small angle detector consisting of two arrays of scintillator counters, called V0A (mexican part) and V0C (french part), which are installed on either side of the ALICE interaction point. This detectors have several functions. They provide minimum-bias triggers for the central barrel detectors in $p+p$ and $A+A$ collisions. These triggers are produced by particles originating from either beam collisions or from secondary interactions in the vacuum chamber elements. Since the dependence between the number of registered particles on the V0 arrays and the number of primary emitted particles is monotonic, V0 serves as an indicator of the centrality of the collision through the recorder multiplicity in the event. Cuts on the number of fired counters and on the total charge can be applied to achieve rough centrality triggers. There are three such triggers: the multiplicity, semi-central and central triggers. In practice and during normal operation, both arrays are required (AND mode) to provide triggers, namely: minimum bias (MB), multiplicity (MT), semi-central (CT1) and central (CT2). An OR mode can also be adopted. In $p+p$ collisions, the efficiency for the detection of at least one charged particle detected in both sides is about 75% when no secondary particle is taken into account. It raises up to 84% when the environment effects are introduced [4, 5]. Interactions of protons with the residual gas of the vacuum chamber generate tracks through the ALICE subdetectors. Minimum bias p-Gas trigger (PG) can help to eliminate false events [6]. Moreover, a large background trigger rate is expected in the muon spectrometer trigger chambers. The absence of MB trigger from V0C alone (OR mode of MB trigger), is a good signal to reject a large part of these false muon triggers. Finally, the V0 detector participates in the measurement of luminosity in $p+p$ collisions with a good precision. The V0 material consists of BC4041 scintillating material (2.5 and 2.0 cm in thickness for V0A and V0C respectively) with 1mm in diameter BCF9929A Wave-Length Shifting (WLS) fibers. The fibers spaced by 1 cm are embedded in the two transverse faces of the segments following the *megatile* technique [7] for the V0A array. There are 32 elementary counters arranged in 4 rings and 8 sectors of 45 degrees. The fibers are grouped by layers of 9 units and glued along the two radial edges of the segments for the V0C array. There are 48 elementary counters of this type which are distributed following two inner rings of 8 counters and two outer rings of 16 counters. The latter ones are connected in groups of two. A picture of each array is shown in figure 1.

The V0A detector is located 340 cm from the vertex on the side opposite to the muon spectrometer whereas V0C is fixed to the front face of the hadronic absorber, 90 cm from the vertex. They cover the pseudo-rapidity ranges $2.8 < \eta < 5.1$ (V0A) and $3.7 < \eta < 1.7$ (V0C) and are segmented into 32 individual counters, each distributed in four rings.

The functionality of the V0 detector has been demonstrated in the first $p+p$ collisions [8] reported by the ALICE collaboration. The time resolution of this detector

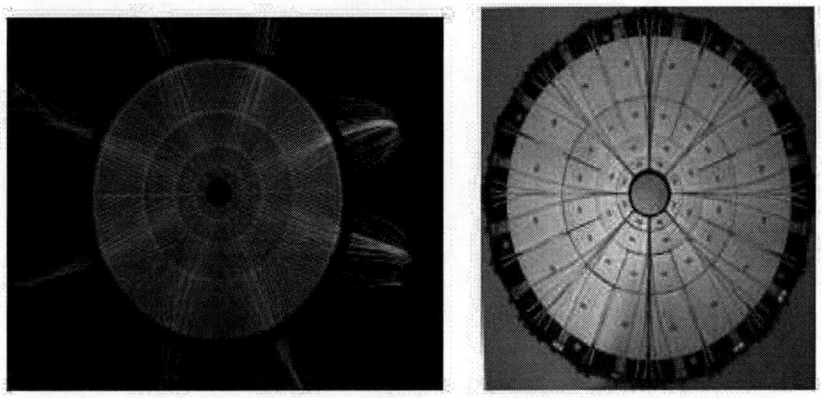

FIGURE 1. Front view of V0A (left) and V0C arrays.

is better than 1 ns. Its response is recorded in a time window of ± 25 nsec around the nominal beam crossing time. For events collected in the first data, the arrival times of particles at the detector relative to this *time zero* is shown in Fig. 2. Note that in general several particles are registered for each event. Particles hitting one of the detectors before the beam crossing have negative arrival times and are typically due to interactions taking place outside the central region of ALICE.

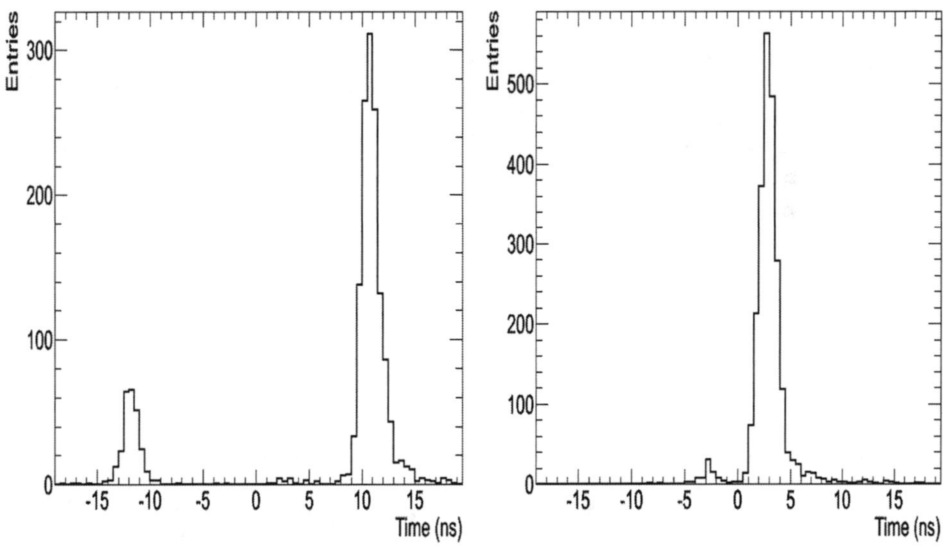

FIGURE 2. Arrival time of particles in the V0 detectors. A number of beam-halo or beam-gas events are visible as a secondary peak in V0A (left) and V0C (right). The figure is taken from Ref. [8].

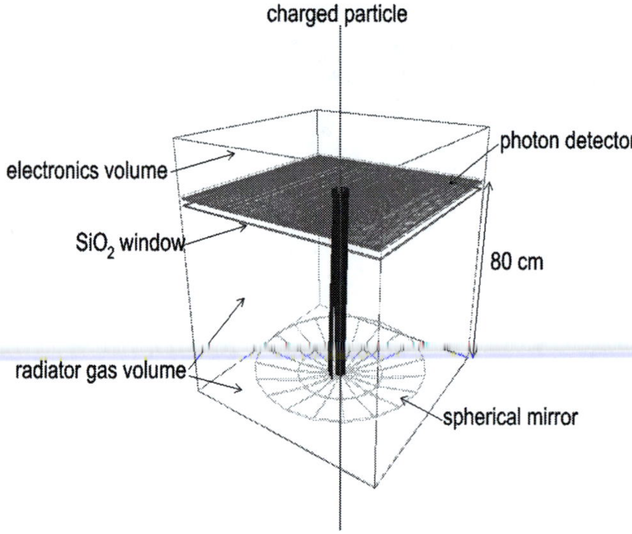

FIGURE 3. First design of the VHMPID prototype.

Another subdetector system where the mexican team contributes is the Very High Momentum Particle Identification (VHMPID). RHIC results have shown the importance of high momentum particles as hard probes and the need for particle identification in a very large momentum range. The ALICE detector has a unique capability to identify a wide variety of particles; however by conception its momentum coverage should be extended to meet new physics challenges at the LHC. The details of the VHMPID upgrade project are reported in the R&D proposal [9]. Figure 3 shows the prototype of one of the modules of the detector.

The approval process, the construction and installation of the new detector may need from three to five years. The proposal to develop and build a full scale prototype is motivated not only by the need to test the new design but also by the possibility to anticipate the contribution to the ALICE physics program and to the TPC calibration, by extending the track-by-track particle identification capability in the high p_t region. In particular, installation at mid-rapidity would be strategic for different reasons: it is the region of the TPC acceptance where tracking and the use of the relativistic rise energy loss could be affected by the central electrode; it is the part of the TPC with the largest drift path to the readout chambers. The simultaneous identification in the TPC and in the VHMPID in a part of the acceptance would open the way to a useful synergy to establish the best possible identification in the high momentum region: in the proton case it should be able to identify in the momentum range of 10 to 26 Gev/c. An early installation will allow for a decision on the full VHMPID based on comparison of performance with the TPC particle identification. The simulation response of the particle identification in the first module which would be of the order of 2 m^2, is shown in figure 4, where a clear identification of the protons up to 25 GeV/c can be seen.

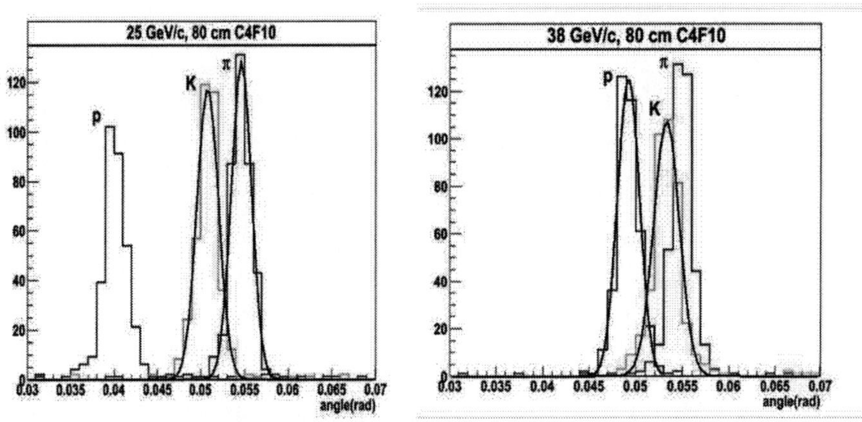

FIGURE 4. Distribution of reconstructed ring Cherenkov angles at 25 (left) and 38 (right) GeV/c after the particle identification algorithm is applied.

ICN-UNAM is participating in physics simulations as well as on R&D of Thick Gas Electron Multiplier (TGEM) detectors.

A cosmic ray detector (ACORDE) is another subdetector system where the mexican collaboration contributes to the ALICE experiment. Details of this detector can be found in A. Fernández' contribution to these proceedings. G. Herrera-Corral is also developing a detector for ALICE, dedicated to study soft difractive physics. A prototype as well as the development of its offline counter part are currently being developed.

POSSIBLE FIRST PHYSICS WITH THE ALICE EXPERIMENT

The argument for studying strange particle production in $A+A$ collisions as evidence of quark-gluon plasma formation is very simple and was suggested in 1982 by Rafelski and Müller [10]. Besides this subject, there are many other variables that could indicate deconfinment, such as multiplicity distributions, collective phenomena in $p+p$ and $A+A$ collision, hadron suppression at high transverse momenta, as well as jet suppression in the away side. Some of these subjects are discussed bellow.

Production mechanism

Production of mesons and baryons depends on the fragmentation processes. While meson production is represented as a simple piece of the string between two $q-\bar{q}$ endpoints, the extension to the case of baryons is not straightforward. Actually there are different scenarios like Diquark, Simple Popcorn, Advanced Popcorn and Baryon Junction production mechanism. A general description of these models can be found in the Pythia [11] manual and references therein. The differences among these models give the possibility to investigate them with data from LHC.

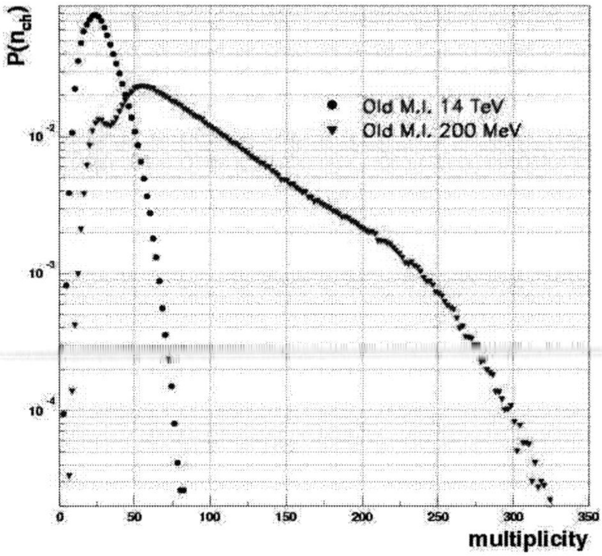

FIGURE 5. Multiplicity distribution obtained with multiple interaction model scaled from 200 GeV up to 14 TeV. The shape looks completely different.

One of the first results to investigate is the multiplicity distribution which could be completely different when it is extrapolated to LHC energies, see Fig. 5. The multiplicity distribution of ALICE at 0.9 TeV has been published [8].

The production rates of baryons and mesons are best exemplified with the ratios of protons to pions, obtained from the different models. Indeed in this ratio there are important differences observed both in the whole pseudorapidity and in the central rapidity ranges, since the multiplicity distributions are different depending on the hadron production mechanism [12]. Using the PYTHIA event generator, it is possible to extract the predictions of the baryon to meson ratio as a function of different multiplicity bins [12]. This and other variables allow us to characterize events and consequently to extract information on the production mechanisms.

From the experimental point of view, the DELPHI collaboration has published [13] results to discriminate between the Diquark and the Popcorn models. However the results are still not completely decisive.

In addition to accounting for the production mechanism, it is necessary to consider the radial flow to understand the experimental results in $p+p$ as well as $A+A$ collisions. Radial flow, discovered [14] in $A+A$ collisions by NA44 experiment, brought new problems to understand the data, since the observed production of protons and pions was different.

This anomalous behavior of the baryon production is called the baryon puzzle and refers to the p/π^+ and $\bar{p}/\pi^-$ ratios measured in $A+A$ collisions and even in $p+p$ collisions.

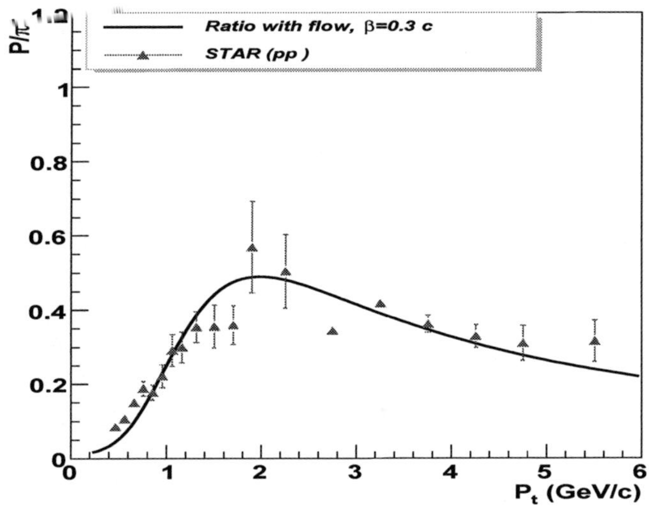

FIGURE 6. Proton to pion ratio (solid line) from $p+p$ collision using HIJING [15], compared to data from STAR [16]. The figure is taken from Ref. [17].

Simulation of $p+p$ collisions allow to extract flow parameters that fit reasonably well the experimental results, both for the p_t spectra and the proton to pion ratio. These parameters are very close to the ones extracted from experiments in a thermal model analysis. Figure 6 shows the proton to pion ratio, where radial flow is taken into account to describe the production of pions and protons from $p+p$ collisions using HIJING [15], compared to experimental data [16]. Details on how to introduce radial flow are discussed in Refs. [17]. The success to qualitatively reproduce the data in a large range of transverse momentum, indicates that the presence of radial flow should be taken into consideration in any attempt to explain the so called baryon puzzle.

Transverse momentum spectra in $A+A$

An alternative way to understand RHIC data, is to consider effects [18] of finite size for the system of emitting particles. In this scenario, particles are considered as confined when their wave functions are constrained to satisfy a given condition at the boundary before freez out. In order to calculate the proton and pion spectra, we calculate their wave functions solving the Dirac and Klein-Gordon equations for protons and pions, respectively, with vanishing wave function at the surface of a sphere of radius R, representing the radius of the overlap region in the collision, see Ref. [18] for details of the calculations.

The spectra thus obtained, can be compared to experimental pion and proton spectra for $Au+Au$ collisions at $\sqrt{s_{NN}} = 200$ GeV [19]. It is possible to show that for temperatures

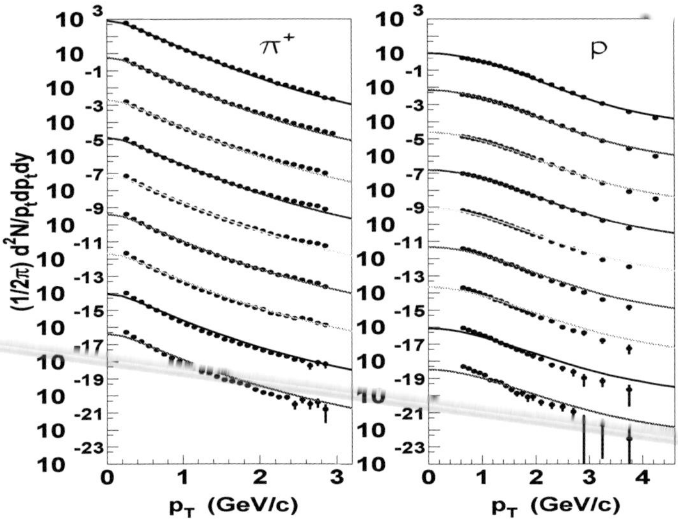

FIGURE 7. Invariant π^+ (left panel) and p (right panel) distributions as function of p_t for different collision centralities, calculated as described in Ref. [18] and compared to PHENIX data.

and collective transverse flow within values corresponding to kinetic freeze-out conditions, the transverse momentum distributions can be described with common values of temperature and expansion velocities for all collision centralities for pions and for most of the collision centralities for protons. Figure 7 shows the transverse momentum spectra for pions and protons at different collision centralities. The upper curves correspond to the most central collisions, (0 - 10)% with the centrality decreasing in intervals of 10% [19] multiplied by successive factors of 10^{-2}. The size of the equivalent spherical region is calculated according to the relation $R = R_0 + C(N_{part}/2)1/3$, with $R_0 = 1$ fm and C = 1.28. The curves represent the theoretical calculation with $T_{p,\pi} = 117$ MeV, $\beta_\pi = 0.6$ and $\beta_p = 0.53$, which describe the most central collisions data. Previous results indicate that finite size of the interaction region and a simple scaling law of this with the number of participants in the collision, is necessary to describe $Au + Au$ data at RHIC energies. It should be interesting to see the behavior at LHC energies.

Event structure in $p + p$ and $A + A$ collisions

Azimuthal correlations studies of high momentum particles have considerably enhanced our understanding of the parton energy loss mechanism in the dense matter created in the aftermath of a heavy ion collision. The two-particle (dihadron) relative azimuthal correlations ($\Delta\phi$) technique, provides an alternative approach for accessing the properties of jets. Two classes of hadrons, trigger and partner hadrons, typically from different p_t ranges, are correlated with each other. Jet properties are extracted on

a statistical basis from the $\Delta\phi$ distribution built of many events. However for a given combination of triggering momenta and associated particles, a double hump structure was observed in the away-side associated particle correlations [20] and/or a broadening of the away-side peak [21].
One way to investigate the shape of the $p+p$ events is through the Event Shape Analysis (ESA), which has been used in e^+e^- collisions and proved to be a useful tool for the discovery of the gluon. This technique was recently successfully applied to the case of $p+p$ collision [22] at RHIC energies. The shape variables are the Thrust (T) and the Recoils (R), defined respectively as:

$$T \equiv \underbrace{max}_{\vec{n}_T} \frac{\Sigma_i |\vec{q}_{\perp,i} \cdot \vec{n}_T|}{\Sigma_i |\vec{q}_{\perp,i}|}, \tag{1}$$

and

$$R \equiv \frac{|\Sigma_i \vec{q}_{\perp i}|}{\Sigma_i |\vec{q}_{\perp i}|}, \tag{2}$$

where the sum runs over all particles in the final state within the acceptance, $\vec{q}_{\perp,i}$ represent the momentum components transverse to the beam and $\vec{n}_T$ is the transverse vector that maximizes the ratio. T is related to the sphericity in the transverse plane, $1-T$, and takes values from 0-1. Events with sphericity close to 0 correspond to a dijet detected within the acceptance. Events with sphericity values close to 0.5 represents isotropic events. ESA reveals that the double hump structure for the away-side jet observed in $A+A$ and not in $p+p$ collisions can thus be an indication of a enhanced probability driven by medium properties. The processes occurring after the hard parton scattering, that in simulations require large intrinsic k_t values, may reflect and enhanced probability for gluon emission with large angles relative to the direction of the parent parton. This result can be understood in simulations by the splitting of jets produced by quarks and gluons. Fig. 8 shows simulated $p+p$ events, splitting them in gluon and quark jets. We can see in the right panel the double hump structure which correspond to a dijet in the away side, while in the left panel we observe the usual distribution of a dijet event, when in reality we have shown that it contain all possible events.
ESA seems to be a powerful tool to investigate other aspects of the events, through the sensitivity of the variables [23] thrust and sphericity to the gluon parton distribution functions. I should also mention that inspired in ESA, it has recently been shown in Ref. [24] that the ratio of processes containing two and one jets in the away side in $A+A$ collisions are enhanced relative to the same ratio of processes in $p+p$ collisions. This effect comes from the fact that the parton with the largest (shortest) trajectory in the medium travels in average a larger (shorter) distance than the parton in a processes with one particle in the away side, thereby losing in average a greater (smaller) amount of energy. This geometrical effect produces in average that one of the two particles in the away side gets absorbed while the other escapes producing in average a double hump structure. This scenario can be tested with the first $Pb+Pb$ data from ALICE.

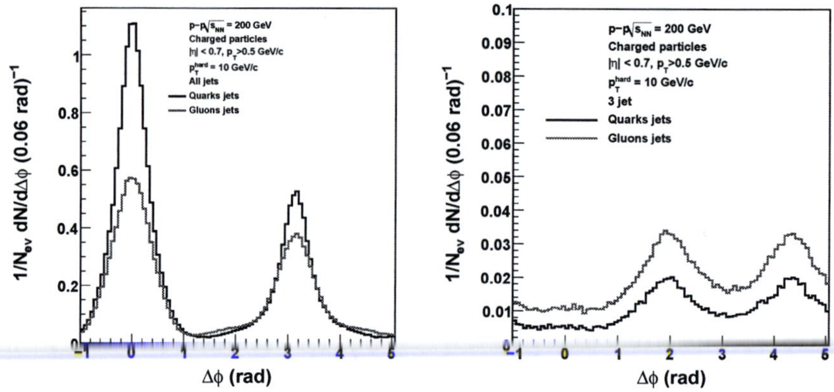

FIGURE 8. Azimuthal correlation splitting the jets coming from fragmentation of quarks and gluons, simulating $p+p$ collisions at $\sqrt{(s_{NN})} = 200$ GeV, and considering processes with transverse momentum larger than 10 GeV/c. The left panel shows the situation when all jet events are considered while the right panel considers the situation with only three jet events.

Conclusion

We have presented a review of the status of the mexican participation in detector construction for the ALICE experiment. Two of the detectors, V0 and ACORDE, are working successfully. ACORDE has been taking data even before the first $p+p$ collisions, and these were used to calibrate other detectors. V0, as a trigger detector, played and important role in the first data taking, helping to discriminate real from beam-gas interaction events. This detector behaved as expected according to the simulations produced at the beginning of its construction [6]. The VHMPID detector is planned as an upgrade for ALICE. Regarding the first possible analysis, I have highlighted some of the recent work on phenomenology and simulation on $p+p$ collisions. The latter can be performed analyzing about a million events. Given the current existence of real data, I should mention that our group is working on analyzing these data, as we speak. Some of these analysis include: hadron production mechanisms, collective phenomena in $p+p$ and in $A+A$ collisions and characterization of the events by means of the shape variables which was successfully applied at RHIC energies.

ACKNOWLEDGMENTS

I am grateful to A. Ayala for his interesting comments. Support for this work has been received by PAPIIT-UNAM under grant number IN116508 and CONACyT.

REFERENCES

1. F. Carminati, et al. (The ALICE Collaboration), *J. Phys. G*: Nucl. Part. Phys. **30** 1517 (2004); B. Alessandro, et al. (The ALICE Collaboration), *J. Phys. G*: Nucl. Part. Phys. **32** 1295 (2006); K. Aamodt, et al. (The ALICE Collaboration), *JINST* **3** S08002 (2008)
2. K. J. Eskola, P. V. Ruuskanen, S. S. Rasanen, and K. Tuominen. *Nucl. Phys.*, **A696**, 715728 (2001).
3. Dmitri Kharzeev, Eugene Levin, and Marzia Nardi. *Nucl. Phys.*, **A747** 609629 (2005).
4. The ALICE Collaboration, ALICE forward detectors: FMD, TO and VO: Technical Design Report, CERN-LHCC-2004-025, http://cdsweb.cern.ch/record/781854.
5. J. Conrad et al., *Minimum bias triggers in proton-proton collisions with VZERO and pixel detectors*, ALICE Internal Note ALICE-INT-2005-025, http://cdsweb.cern.ch/record/960438.
6. E. Cuautle et al., *Rejection of beam gas interactions in pp collisions and timing requirements*, ALICE Internal Note ALICE-INT-2004-021, https://edms.cern.ch/document/494149/1.
7. J.R. Alfaro et al., *Simulation of the V0A detector*, ALICE Internal Note ALICE-INT-2006-018, https://edms.cern.ch/document/773744/1.
8. K. Aamondt, et al., (The ALICE Collaboration) *Eur. Phys. J.* **C65**, 111 (2010).
9. A. Dimauro, et al., Proposal for a Very High Momentum Particle Identification (VHMPID) R&D project, ALICE Internal Note ALICE-2009-032, https://edms.cern.ch/nav/ALI-0000000250/ALI-0000000255
10. J. Rafelski and B. Müller, *Phys. Rev. Lett.*, **48**,1066 (1982) Bonner.- B.E. Bonner et al., *Phys. Rev.* **D 38**, 729 (1988).
11. T. Sjöstrand et. al., *PYTHIA 6.3 physics and manual* (hep-ph/0308153).
12. E. Cuautle and G. Paić, *Study of the pion and proton production in pp collisions at 14 TeV* ALICE-INT-2005-027, https://edms.cern.ch/document/673703/1
13. P. Abreu, et al. (DELPHI Collaboration), *Phys. Lett.* **B490**, 61 (2000)
14. I. G. Bearden, et al. (NA44 Collaboration), *Phys. Rev. Lett.* **78**, 2080 (1997)
15. Miklos Gyulassy, Xin-Nian Wang, *Comput. Phys. Commun.* **83** (1994) 307.
16. J. Adams et al. *Phys. Lett.* **B637** 161 (2006).
17. E. Cuautle and G. Paić, *J.Phys.* **G35**, 075103,(2008).
18. A. Ayala, E. Cuautle, J. Magnin, L. M. Montaño *Phys. Rev.* **C74**, 064903 (2006); A. Ayala, E. Cuautle, J. Magnin, L. M. Montaño, A. Raya *Phys. Lett.* **B634**, 200 (2006).
19. S.S. Adler et al. (PHENIX Collaboration), *Phys. Rev.* **C 69**, 034909 (2004).
20. A. Adare, et al. (PHENIX Collaboration), *Phys. Rev.* **C 78**, 014901 (2008).
21. Wang F. et al. (STAR Collaboration), *Nucl. Phys.* **A 774**, 129 (2006).
22. A. Ayala, E. Cuautle, I. Domínguez, A. Ortiz, G. Paić *Eur. Phys. J.* **C62**, 535 (2009); A. Ortiz and G. Paić, *Event Shape Analysis in ALICE* ALICE-INT-2009-015, https://edms.cern.ch/document/1019980/1
23. E. Cuautle, G. Paić and A. Ortiz, *Sensitivity of the parameters measured in pp collisions on the gluon PDF*, arXiv:1001.0613 [hep-ph]
24. A. Ayala, J. Jalilian-Marian, J. Magnin, A. Ortiz, G. Paić and M. E. Tejeda-Yeomans, *Phys. Rev. Lett.* **104**, 042301 (2010).

B Physics at the D0 experiment
A Mexican review

E. De La Cruz-Burelo

Department of Physics,
Centro de Investigaciones y de Estudios Avanzados
del Instituto Politécnico Nacional,
Av. IPN 2508, México D.F., México.

Abstract.
On April of 1992 a Mexican group from Cinvestav officially joined the D0 experiment, one of the two experiments in the Tevatron collider at Fermilab. The seed for this experimental group on high energy physics from Cinvestav was planted in Mexico in some measure by Augusto Garcia, to whom this workshop is in memorial. Augusto's efforts and support to groups dedicated to this area was clear and important. Some of these seeds have given origin to today's established Mexican groups on experimental high energy physics, one example of this is the Mexican group at D0. I present here a short review of some of the D0 results on which the Mexican group has contributed, emphasizing the last decade, which I have witnessed

Keywords: Bottom baryons
PACS: 14.20.Mr, 14.40.Nd, 13.30.Eg, 13.25.Hw

INTRODUCTION

I have the opportunity to talk to Augusto in many occasions and in different periods of my academic life: as an undergrad student, as a graduate student, and finally as a colleague. Augusto had a very strong personality with well defined thoughts, a person with whom you can spend hours talking about physics, most of the time it was him doing the talking, but you were always with the feeling of be learning a lot each time you have the pleasure of share some time with him. I was deeply sorry the moment I learned that I could not have another chat with him or to hear his voice talking to somebody in the physics department at Cinvestav.

When I was invited to give a presentation in this whorshop on Augusto's memorial, I decided to present something which I discussed some time ago with Augusto and what I think is almost unknown to most of the part of the small community of high energy physics in Mexico, this is, how the Mexican group in the D0 experiment at Fermilab evolved during the last decade to become a strong group with a very well defined research line which identifies it in the international community of high energy physics.

The Mexican group was officially accepted in 1992 by D0, it was integrated at the beginning by two faculties, but it reduced to one faculty and students shortly after. For Run II of the Tevatron one of this students became the second Mexican faculty at D0. I myself became the the third faculty in the group couple of years ago. At the time I joined the group in 2001 as graduate student, the group was integrated by two faculties and two

CP1259, *First CINVESTAV-UNAM Symposium on High Energy Physics*
edited by J. C. D'Olivo, R. López-Fernández, and M. Á. Pérez

students, including myself. The number of students in the group has oscillated with the years, reaching a maximum of eight students in 2008. Currently the group counts with three faculties and six students. Since the Mexican group joined D0, seven students have obtained the Master in Science degree, and eight students the degree of Doctor in Science, all of them with their work at the D0 experiment.

The Mexican group concentrated in Run II of the Tevatron on the study of *B* hadrons, the physics of particles which heaviest quark is a *b*. For this, the Mexican group joined the *B* physics group at D0, this group was new in RunII in the sense that the D0 detector has a new tracking and vexterxing systems to determine the production and decay point of particles. Since the early measurements the Mexican group contributed heavily to these efforts, and today still continues producing excellent measurements on this subject. In this contribution I will first present some of the early measurements in the *B* physics group, then I will present some of the more recent measurements, and I will conclude with some prospects on the natural evolution of this group to the LHC.

THE D0 DETECTOR

The D0 detector is described in detail elsewhere [1]. The components most relevant to the analyses presented here are the central tracking system and the muon spectrometer. The central tracking system consists of a silicon microstrip tracker (SMT) and a central fiber tracker (CFT) that are surrounded by a 2 T superconducting solenoid. The SMT is optimized for tracking and vertexing for the pseudorapidity region $|\eta| < 3$ ($\eta = -\ln[\tan(\theta/2)]$ and θ is the polar angle) while the CFT has coverage for $|\eta| < 2$. Liquid-argon and uranium calorimeters in a central and two end-cap cryostats cover the pseudorapidity region $|\eta| < 4.2$. The muon spectrometer is located outside the calorimeter and covers the pseudorapidity region $|\eta| < 2$. It comprises a layer of drift tubes and scintillator trigger counters in front of 1.8 T iron toroids followed by two similar layers behind the toroids.

B HADRONS LIFETIME MEASUREMENTS

The *B* Physics program at D0 was very rich and ambitious since the beginning of RunII of the Tevatron. There were many *B* hadron properties to measure, many of them unknown at the time that Run II of the Tevatron started. Lifetimes measurements which could be compared directly with predictions were one of many subjects that the group expected to cover and to perform many new measurements. However, specially for lifetime measurements, the *B* physics group had first to prove to the scientific community that the new tracking and vextex detectors were working correctly and well understood, and that we could reach precisions comparable to previous measurements. The Mexican group got involved in this subject from the beginning, participating on many of the lifetime measurements at D0, and at the same time producing many tools and new techniques which are now widely used in the *B* Physics group at D0.

In this section I first present a very short introduction to the technique for lifetime measurements used at D0, and then I show some of the measurements performed by the

B physics group with a strong Mexican participation in the extraction of the property from data. The first measurement presented, the average lifetime of all B hadrons helped the group to show that the new vertex and tracking detectors, as well as the montecarlo simulation of B decays, were working correctly and that they were well understood. The second measurement, the B^+ lifetime, was the first lifetime measurement of a fully reconstructed B hadron performed at D0. It allowed us to measure a property very well known by the scientific community, from which we could compare our detector resolutions and precisions to other experiments, and it also allowed us to show that we were able to reproduce previous measurements with very competitive precisions. Finally, I present two more measurements on the B_s and the Λ_b lifetimes. Both measurements were PhD. thesis work of students graduated from the group and published results from D0.

Technique for lifetime measurements

The decay time of a B hadron can be determined by measuring the distance traveled by the hadron in a plane transverse to the beam direction, and then applying a correction for the Lorentz boost. The transverse decay length is defined as $L_{xy} = \boldsymbol{L}_{xy} \cdot \boldsymbol{p}_T / p_T$ where $\boldsymbol{L}_{xy}$ is the vector that points from the production vertex to the decay vertex of the particle, and $\boldsymbol{p}_T$ is its transverse momentum. The event-by-event value of the B hadron proper transverse decay length, λ, is given by:

$$\lambda = \frac{L_{xy}}{(\beta\gamma)_T^B} = L_{xy} \frac{cM_B}{p_T}, \tag{1}$$

where $(\beta\gamma)_T^B$ and M_B are the transverse boost and the mass of the B hadron.

Sometimes it is not possible to fully reconstruct the B hadron, so not the mass, the transversal momentum, or the decay vertex are known. However, if it is possible to identify the decay vertex using another particle which comes from the decay vertex of the B hadron, it is still possible to use eq. 1, but it is necessary to include a correction factor to account for the part of the B hadron boost which is not known. Examples of these are the decays of B hadrons of the form $B \to J/\psi(\mu^+\mu^-) + X$, in which the decay vertex of $J/\psi \to \mu^+\mu^-$ can be considered equal to the decay vertex of the B hadron, this due to the very short lifetime of the J/ψ. For this case we can re-write eq. 1 as:

$$\lambda = \frac{L_{xy}}{(\beta\gamma)_T^B} = L_{xy} \frac{cM_{J/\psi}}{p_T^{J/\psi} F(p_T^{J/\psi})}, \tag{2}$$

where $F(p_T^{J/\psi})$ is the average value of the ratio of proper decay lengths $\lambda_{J/\psi}/\lambda_B$ as funtion of the transversal momentum of the J/ψ, here denoted by $p_T^{J/\psi}$. The values of F has to be estimated from montecarlo simulation.

The proper decay length distribution coming from any particle decay can be modeled by a exponential decay convoluted with a resolution function to take into account the detector smearing effects. The decay constant of the exponential decay is the average value of the lifetime of the particle. By fitting techniques like maximum likelihood or χ^2 it is possible to extract this decay constant from data, and in this way to measure the lifetime of B hadrons from the distribution of proper decay length.

Inclusive B hadrons lifetime

At the begining of Run II of the Tevatron, there was not enought data to fully reconstruct B hadrons, then the first measurement done in the B Physics group of D0 was the average of the lifetimes of all B hadrons [2, 3] contained in a sample of decays $J/\psi \rightarrow \mu^+\mu^-$. The main idea is that in this sample there are mainly two components, one -which is the dominant- of J/ψ coming from direct production at the collision, this is $p\bar{p} \rightarrow J/\psi(\mu^+\mu^-) + X$, an other coming from B hadrons, $p\bar{p} \rightarrow B \rightarrow J/\psi(\mu^+\mu^-) + X$. Due to the very short lifetime of the J/ψ, consistent with zero compared to our detector resolution, the decay vertex of $J/\psi \rightarrow \mu^+\mu^-$ can be considered the same as the decay vertex of the B hadron, and by using a corrrection factor as indicated in the previous section, it is possible to determine the proper decay length of the each B hadron candidate. Although this measurement was not meant for publication, besides proceedings of important conferences on high energy physics, it was a very important measurement for the B physics group and the collaboration. It represented the first measurement in which the new vertex and tracking detectors were used, and also the tools of montecarlo simulation played an important role in the measurement. Also the triggers system, tracking reconstruction software, vertex reconstruction software, and many important parts of the framework of the collaboration were tested with this measurement. The success of the experiment performing this first measurement was presented in many conferences as a proof of the good path the experiment was taking since the begining of the Run II of the Tevatron.

The Mexican group participated in many stages of this measurement [3], from the data selection to the final measurement. Figure 1 shows the distribution of the proper decay length extracted from data and the correction factor used in the measurement of the average lifetime of all B hadrons. From approximately 40pb^{-1} of data, the average lifetime of all B hadrons was determined to be [2, 3]:

$$c\tau = 468 \pm 7(stat.) \pm 22(syst.)\mu m \tag{3}$$

Under some assumptions [4], this averaged B hadrons lifetime can be used to estimate the value of $|V_{cb}|$, an element of the CKM matrix.

Lifetime of B^+

The B^+ is the lightest and the most produced B hadron. It is formed by the combination of a b and an u quarks. As expected, the B^+ was the first B hadron to be observed in the data collected by D0, and its lifetime measurement was the first the B physics group performed on a fully reconstructed B. For the Mexican group it was natural to extend its new adquired expertise in the measurement of the average inclusive B hadrons life-

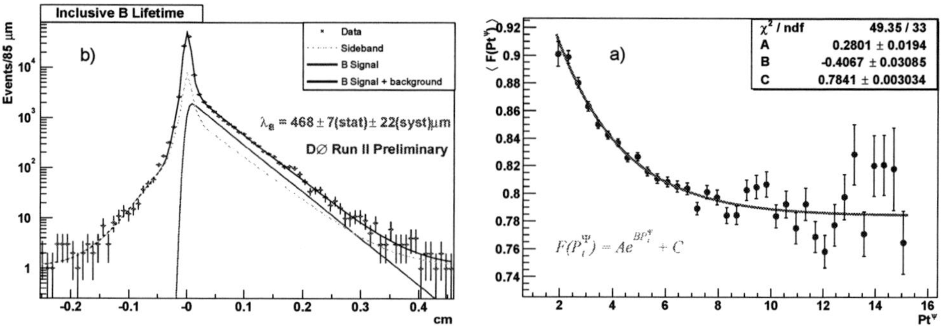

FIGURE 1. (Right) Proper decay length distribution of $J/\psi(\mu^+\mu^-)+X$ events with fit results superimposed. (Left) Correction factor $F(p_T^{J/\psi})$ extracted from montecarlo simulation and used in the measurement of the average lifetime of all B hadrons.

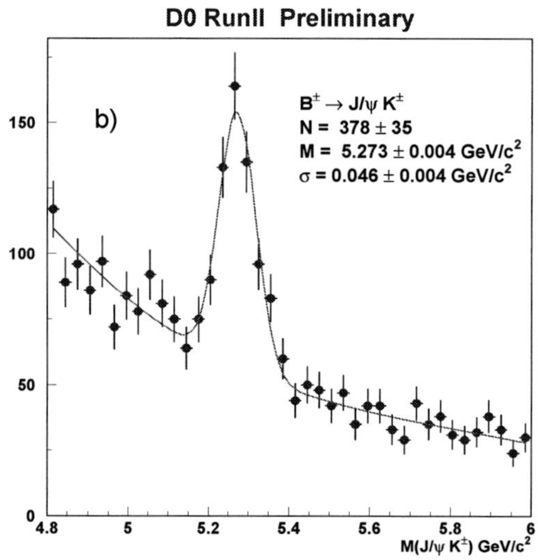

FIGURE 2. $B^{\pm}$ reconstructed in the decay channel $B^{\pm} \to J/\psi + K^{\pm}$.

time to cover the measurement of the B^+ lifetime. This property of the B^+ was known with great precision due to measurements from B factories, in adition, it was possible to compare this measurement with predictions. All this makes this lifetime measurement a great opportunity to test our measurement techniques and detector, as well as our alignment and calibration methods. As in the measurement of the average inclusive B hadrons lifetime, the Mexican group actively participated in all stages of this measurement [5]. In the decay channel $B^{\pm} \to J/\psi + K^{\pm}$, from approximately 40pb^{-1} of data, the lifetime of the $B^{\pm}$ hadron was determined to be [2, 5]:

$$\tau_{B^{\pm}} = 1.761 \pm 0.24(stat.)ps \tag{4}$$

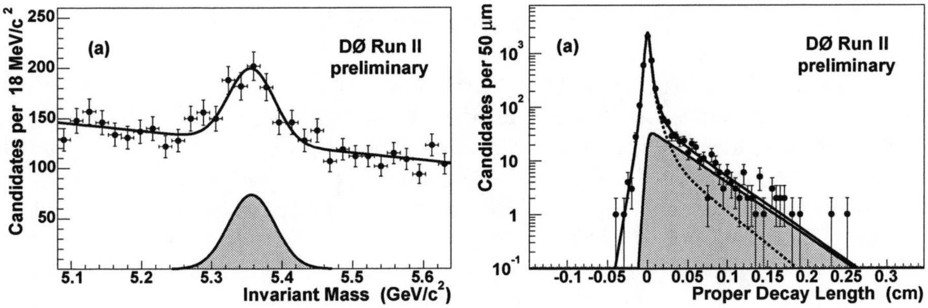

FIGURE 3. Invariant mass distribution for B_s candidates (left) and its proper decay length distribution (right) with the fit results superimposed.

This lifetime was in very good agreement with previous measurements and predictions.

B_s lifetime measurment in $B_s \to J/\psi\phi$

With the accumulation of data collected by the D0 experiment, the Mexican group as part of the B_physics group at D0, focused on reconstructing other B hadrons like the $B^0(\bar{b}d)$, $B_s(\bar{b}s)$, and $\Lambda_b(bdu)$. The B_s system was of great interes due to the possibility of observing CP violation in it, and the idea of use this as test in the search for physics beyond de Standard Model of particles. The Mexican group followed this research path from the beginning, performing the first measurement in D0 of the average B_s lifetime [6]. The B_s was reconstructed in the decay channel $B_s \to J/\psi\phi$ and in approximately 220 pb^{-1} of data, the lifetime of the B_s was found to be [7, 6]:

$$\tau(B_s) = 1.444^{+0.098}_{-0.090}(stat.) \pm 0.020(syst.)ps. \tag{5}$$

Simultaneously, we measured the lifetime of the B^0 in the decay $B^0 \to J/\psi K^*$:

$$\tau(B^0) = 1.473^{+0.052}_{-0.050}(stat.) \pm 0.023(syst.)ps. \tag{6}$$

These was combined to determine the ratio of lifetimes:

$$\frac{\tau(B_s)}{\tau(B^0)} = 0.980^{+0.076}_{-0.071}(stat.) \pm 0.003(syst.) \tag{7}$$

At the moment of its publication, this measurement of the B_s lifetime represented the most precise measurement of a single experiment, and the ratio of lifetimes was in good agreement with QCD models based on heavy quark expansion [8].

The Mexican group continues actively working on the B_s system, specially on the determination of the CP violation phase and its implication on physics beyond the Standar Model of particles.

Λ_b lifetime measurement in $\Lambda_b \to J/\psi\Lambda$

As data was being accumulated, it was possible to reach the observation of more heavier particles other than the B mesons. The lightest B baryon, and the only one observed before the Run II of the Tevatron, was the Λ_b. However, the small statistics accumulated by previous experiments was easily surpassed by the statistics in D0. This increase on the statistics of Λ_b particles allowed us to improve the experimental precision on the Λ_b lifetime, and in addition there was a great interest on this lifetime due to an earlier difference between experiments and prediction which was named as the Λ_b *lifetime puzzle*[9]. Recent theoretical calculations of the ratio of lifetimes $\tau(\Lambda_b)/\tau(B^0)$, which include next-to-leading order effects in QCD [10], corrections at $\mathcal{O}(1/m_b^4)$ in HQET [11], and lattice QCD studies [12], have led to a prediction of $\tau(\Lambda_b)/\tau(B^0) = 0.88 \pm 0.05$ [13], that significantly reduced this difference ($\tau(\Lambda_b)/\tau(B^0) = 0.800 \pm 0.053$ was the world average in 2004 [14]). However, a recent precise measurement [15] by CDF experiment reports a value of the Λ_b lifetime consistent with b meson lifetimes, and the ratio $\tau(\Lambda_b)/\tau(B^0)$ consistent with unity. This measurement, which statistically dominates the current world average, has turned around the Λ_b lifetime puzzle, by making the prediction to be lower than the experimental measurement.

At D0 we have measured the Λ_b lifetime in the decay channel $\Lambda_b \to J/\psi\Lambda$ in two different ocassions. First in 2005 [16, 17] followed by an update with more statistics in 2007. The Mexican group participated in all the process of the 2005 measurement, and ex-members of the group participated in the 2007 update. Due to the similarity of the decay channels, simultaneously to the Λ_b lifetime, the B^0 lifetime was measured in the decay channel $B^0 \to J\psi(\mu^+\mu^-)K_S^0(\pi^+\pi^-)$. The results of the most recent measurement of the Λ_b and B^0 lifetimes can be summarized as:

$$\begin{aligned} c\tau(\Lambda_b) &= 365.1^{+39.1}_{-34.7}\,(\text{stat}) \pm 12.7\,(\text{syst})\ \mu\text{m}, \\ c\tau(B^0) &= 450.0^{+23.5}_{-22.1}\,(\text{stat}) \pm 14.9\,(\text{syst})\ \mu\text{m}, \end{aligned} \tag{8}$$

from which:

$$\begin{aligned} \tau(\Lambda_b) &= 1.218^{+0.130}_{-0.115}\,(\text{stat}) \pm 0.042\,(\text{syst})\ \text{ps}, \\ \tau(B^0) &= 1.501^{+0.078}_{-0.074}\,(\text{stat}) \pm 0.050\,(\text{syst})\ \text{ps}. \end{aligned} \tag{9}$$

These can be combined to determine the ratio of lifetimes:

$$\frac{\tau(\Lambda_b)}{\tau(B^0)} = 0.811^{+0.096}_{-0.087}\,(\text{stat}) \pm 0.034\,(\text{syst}), \tag{10}$$

Figures 4 and 5 show the mass and transverse proper decay length distributions for the Λ_b and B^0 candidates. Likelihood fit results are superimposed. This measurement found the lifetime to be consistent with the world average [19], and the ratio of Λ_b to B^0 lifetimes consistent with the most recent theoretical predictions [13]. More statistics is needed to solve the Λ_b lifetime puzzle, and the Mexican group continues adding more data to this measurement to improve the precision of the experimental value in order to give a definitive word on this puzzle.

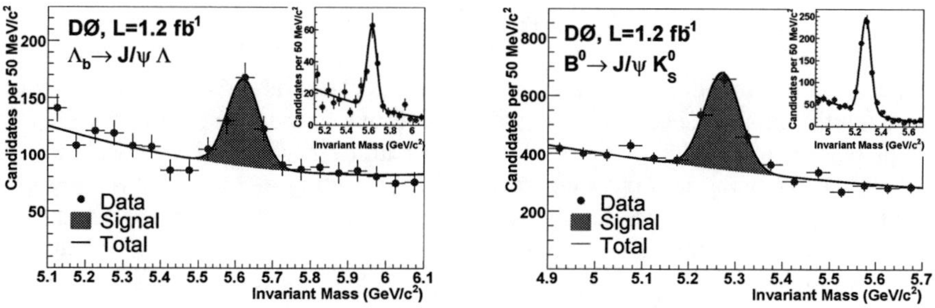

FIGURE 4. Invariant mass distribution for Λ_b (left) and B^0 (right) candidates with the fit results superimposed. The inserts show the mass distributions after requiring $\lambda/\sigma_\lambda > 5$.

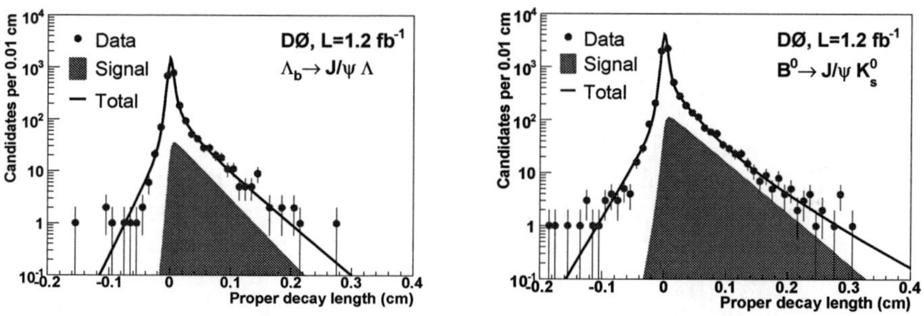

FIGURE 5. Proper decay length distribution for Λ_b(left) and B^0(right) candidates, with the fit results superimposed. The shaded region represents the signal.

NEXT STEP: THE LHC

With the starting of collisions at the Large Hadron Collider (LHC) this year (2010), it is natural that most of the groups of experimental high energy physics in the world will migrate to experiments at the LHC. This is also the case for the Mexican group which started many years ago at D0, and that now it is also part of the Compact Muon Solenoid experiment (CMS) at the LHC. *B* physics will be the earlier physics to be done at the LHC, and it is there where the Mexican group could take adventage of its expertise, and it could takes the opportunity to prepare the natural evolution of the group to the study of other subjects which will be possible to attack at the LHC, in addition to *B* physics. This new beginning at CMS will be however different to how it was in the D0 experiment. The group has an excellent expertise on bringing to end many different an complicated studies on *B* physics and other subjects, plus that it is a well established group in Mexico, with many of the students which graduated in the group being faculties in different institutions in the country. In addition, each year the group attract many new students which will become its strengh in CMS.

SUMMARY

During the last decade, the Mexican group at D0 has participated directly in many more measurements than those presented here. This without taking into account those in which the group has participated in an indirectly way, as part of the internal review committees, as part of the review in the B physics group, or by providing tools and techniques which are used by other groups. Also, many other measurements have came to an end thanks to ex-members of the Mexican group which continue their work in D0 after leaving the group. In addition to the analyses presented here, other analyses in which members or ex-members of the group have directly participated are:

- $X(3872)$ observation and properties [20].
- B_s lifetime in semileptonic decays [21].
- B^0 and B_s lifetimes and decay amplitudes in $B^0 \to J/\psi K^*$ and $B_s \to J/\psi\phi$ [22].
- Lifetime difference and CP-violating phase in the B_s system [23].
- Ξ_b^- observation [24].
- Ω_b^- observation [25].

I am probably forgetting other analyses in which the group has worked during all these years, this is not meant to be an extensive list, it is just with the idea of showing how that seed planted by Augusto many years ago has given place to a group which is now very well stablished and which has contributed with the preparation of a very good part of the Mexican physicist working today on experimental high energy physics in Mexico. During our talks on the evolution of the Mexican group at D0, I could see that Augusto was happy to see how that seed he planted is more like a tree now, but he continued to ask for more, he asked to become a forrest, and in a future to prepare our own experiments in our country, in Mexico. I think many members of the high energy physics community in Mexico share Augusto's feelings on this subject, there is nothing more left to us by Augusto that to work for this to become true some time in the future.

ACKNOWLEDGMENTS

I would like to thanks the organizers of the workshop for the invitation to participate and for an excellent organization. Also I want to acknowledge that as in any collaboration of high energy physics, without the hard work and dedication of all of our colleagues, none of the measurements presented here could have been possible. It is the hard work and dedication of all the members of an experiment what makes possible its succeed.

REFERENCES

1. V.M. Abazov *et al.* (D0 Collaboration), Nucl. Instrum. Methods A **565**, 463 (2006).
2. F. Filthout, *B Physics at D0*; hep-ex/0305062.
3. E. De La Cruz-Burelo, *Inclusive B lifetime at D0*, talk at the April 2003 meeting of the American Physical Society, http://www-d0.fnal.gov/d0pubs/sbdata/2003/030405-DELACRUZBURELO_E-

talk.pdf; E. De La Cruz-Burelo, *Medición del tiempo de vida de los hadrones B usando estados finales con J/ψ*, M.S. Thesis, 2002.
4. S. Stone, *B decays*, pag. 23, 1992 by World Scientific Publishing Co. Pte. Ltd.
5. P.L.M. Podesta-Lerma, *Full reconstruction of B mesons and the B^+ Lifetime at D0*, talk at the April 2003 meeting of the American Physical Society, http://www-d0.fnal.gov/d0pubs/sbdata/2003/030405-PODESTA-LERMA_PLM-talk.pdf.
6. P.L.M. Podesta-Lerma, *Medición del tiempo de vida media del meson B_s en el experimento DZero*, PhD. Thesis 2005.
7. V.M. Abazov *et al.* (D0 Collaboration), Phys. Rev. Lett. **94**, 042001 (2005).
8. I.I.Y. Bigi *et al.*, Phys. Lett. B **293**, 430 (1992).
9. F. Gabbiani *et al.* Phys. Rev. Lett. **D** 68, (2003) 114006-1-4.
10. E. Franco *et al.*, Nucl. Phys. B **633**, 212 (2002).
11. F. Gabbiani, A. I. Onishchenko, A. A. Petrov, Phys. Rev. D **68**, 114006 (2003).
12. M. Di Pierro *et al.*, Phys. Lett. B **468**, 143 (1999).
13. C. Tarantino, Nucl. Phys. B **156**, (Proc. Suppl.), 33 (2006).
14. S. Eidelman et al. (PDG 2004), Phys. Lett. B **592**, 1 (2004).
15. A. Abulencia *et al.*, Phys. Rev. Lett. **98**, 122001 (2007).
16. V.M. Abazov *et al.* (D0 Collaboration), Phys. Rev. Lett. **94**, 102001 (2005).
17. E. De La Cruz-Burelo, *Measurement of the Λ_b lifetime in the decay channel $\Lambda_b \to J/\psi\Lambda^0$ in the D0 experiment at the Tevatron Collider*, PhD. Thesis, 2005.
18. V.M. Abazov *et al.* (D0 Collaboration), Phys. Rev. Lett. **99**, 142001 (2007).
19. W.-M. Yao *et al.*, Journal of Physics G **33**, 1 (2006).
20. V.M. Abazov *et al.* (D0 Collaboration), Phys. Rev. Lett. **93**, 162002 (2004).
21. V.M. Abazov *et al.* (D0 Collaboration), Phys. Rev. Lett. **97**, 241801 (2006).
22. V.M. Abazov *et al.* (D0 Collaboration), Phys. Rev. Lett. **102**, 032001 (2009).
23. V.M. Abazov *et al.* (D0 Collaboration), Phys. Rev. Lett. **98**, 121801 (2007).
24. V.M. Abazov *et al.* (D0 Collaboration), Phys. Rev. Lett. **99**, 052001 (2007).
25. V.M. Abazov *et al.* (D0 Collaboration), Phys. Rev. Lett. **101**, 232002 (2008).

A New Proposal to Measure $K^+ \rightarrow \pi^+ \nu \bar{\nu}$

D. A. Jensen

Fermi National Laboratory PO Box 510, Batavia IL, 60510

Abstract. Do not replace the word "abstract," but do replace the rest of this text. Text should remain 10 pt. If you must insert a hard line break, please use Shift+Enter rather than just tapping your "Enter" key. You may want to print this page and refer to it as a style sample before you begin working on your paper. For an experimentalist in high energy physics, the most interesting experiments that one should pursue is often well known. The very rare CP violating decay of the kaon into a pion and two neutrinos (both the charged and neutral modes) is such an experiment. We have recently proposed the charged kaon experiment to Fermilab as proposal P966. We present her a general overview of that proposal. The goal is to obtain of order 1000 events in order to confront the expected theoretical calculations and to search for physics beyond the standard model.

Keywords: K decay, CP violation.
PACS: 10.14.40.Dx, 20.29.30.Aj, 20.29.40.Gx

INTRODUCTION

It is often the case that very interesting questions in physics are well known by the community. This is particularly the case in high energy physics. A very nice example is in chapter 19 of Okun's[1] first book on particle physics: there is a list of experiments that he suggests should be done, which includes, among others:

- "Check of CP invariance."
- "Check the hypothesis that the muon neutrino and electron neutrino are different particles"
- "Search for the intermediate vector W-boson."

These were, in the 60's, what one might refer to as 'textbook questions'. There was little doubt that these questions should be addressed experimentally. As will be recognized, most of these questions were addressed in succeeding years, gaining considerable accolades from the particle physics community. It is usually the case that these experiments became possible through improvements in experimental and / or accelerator techniques.

The measurement of the branching ratio of the decay $K \rightarrow \pi \nu \bar{\nu}$, where the K and π can be either + or neutral, are currently such experiments. Considerable theoretical and experimental work has been done to understand these processes. The

CP1259, *First CINVESTAV-UNAM Symposium on High Energy Physics*
edited by J. C. D'Olivo, R. López-Fernández, and M. Á. Pérez

recommendation of recent panels[2] emphasizes that the goal of the experiments should be to confront theory at the level of accuracy of the theory. As will be discussed below such an experiment would need to obtain of order 1000 events, as the theory is expected to approach an accuracy of order 3% within the next few years.

In proposal P996[3] as submitted to Fermi National Accelerator Laboratory, we propose to carry out this experiment for the K^+ decay. The technological innovation that makes it possible to better previous experimental efforts by of order 1000 is the idea of using the Tevatron superconducting synchrotron as a stretcher ring for main injector beam. This makes it possible to obtain a large number of K^+'s stopping in a target with minimal instantaneous rate. The detector to be used is an upgrade of a previous detector – the detector used in the Brookhaven National Laboratory Experiment E 987[3].

In the following sections, a broad overview of the proposed experiment is presented. Many details are not included here, but may be found the in the proposal. In particular, cost and schedule, where the experiment might be located and exactly which magnet might be used are not discussed. Accelerator operation and details of the detector upgrades are only briefly sketched.

Theory of $K^+ \rightarrow \pi^+ \nu \bar{\nu}$

The K decays to pions plus neutrinos are particularly interesting because decays proceed at lowest order through loop diagrams, usually highly suppressed with respect to other channels. The lowest order diagrams for the K^+ decay are shown in figure 1. The basic interaction is simple: $\bar{s} \rightarrow d$ through $(\bar{s}_L \gamma^\mu d_L)(\bar{\nu}_L \gamma_\mu \nu_L)$.

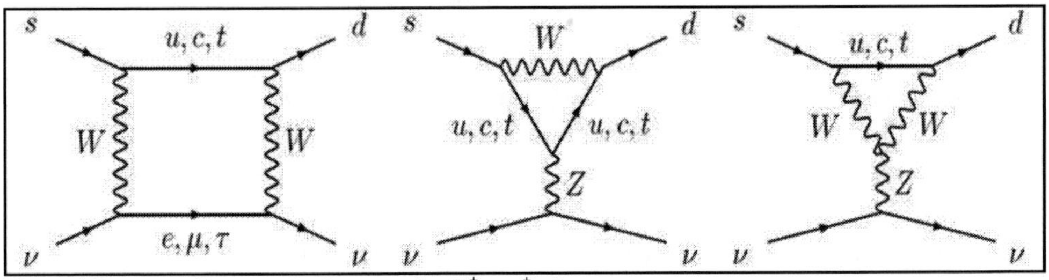

Figure 1 Feynman diagrams for lowest order $K^+ \rightarrow \pi^+ \nu \bar{\nu}$

The branching ratio can then be calculated as[4]:

$$BR(K^+ \rightarrow \pi^+ \upsilon \bar{\upsilon}) = \frac{\tau_{K^+} M_{K^+}^5}{32\pi^3}(1+\Delta_{EM}) \mid f_+^{K^+\pi^+}(0) \mid I_\upsilon^+ \mid \frac{G_F \alpha(M_Z)}{2\pi\sqrt{2}\sin^2\theta_W} Y \mid^2 \quad (1)$$

Where the matrix element Y is

$$Y = V_{td}^{*}V_{ts}X(x_t) + V_{cd}^{*}V_{cx}[\, X(x_c) + |V_{us}|^4\delta P_{cu}\,] \tag{2}$$

$$X(x_q) = \frac{1}{3}\sum_{l} X(x_q, x_l) \tag{3}$$

Where V denotes the CKM matrix, $x_q = m_q^2/M_W^2$ and . $x_l = m_l^2/M_W^2$ The functions X are calculated in perturbation theory.

There are of course many higher order diagrams. Currently, the predicted branching ratio for $K^+\to\pi^+ \nu \bar{\nu}$ is $(8.5 \pm 0.7) \times 10^{-13}$. The theory is expected to improve to an accuracy of order 5% in the next few years. The figure 2 shows the current contributions to the errors in the theoretical calculations. Clearly, this is a very complex subject, and there are many contributions the error on the BR calculation.

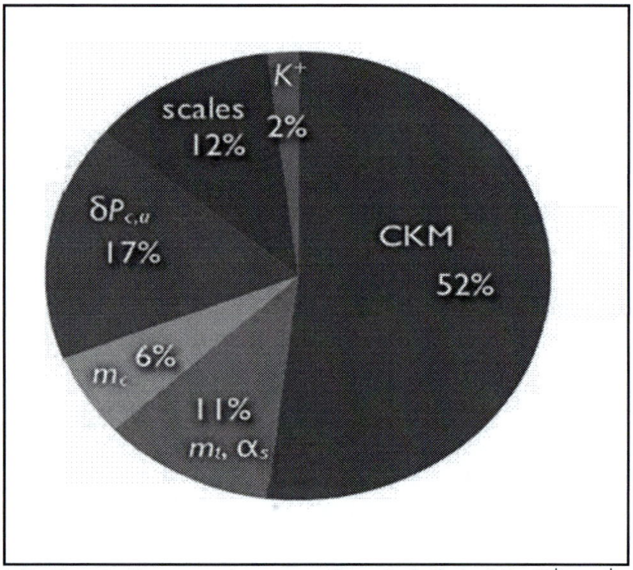

Figure 2 Contributions to the current error on the calculation of the $K^+ \to \pi^+ \nu \bar{\nu}$ branching ratio. The theory as well as the accuracy of the CKM parameters are expect to improve by the time P996 would have results.

It is interesting to note that the theoretical estimation of the branching ration for $K^+ \to\pi^+ \nu \bar{\nu}$ is currently, and is expected to remain, more accurate than the calculation of the branching ration for $K^0 \to\pi^0 \nu \bar{\nu}$.
This sets the level of experimental determination that is of interest. A 3% accuracy requires of order 1000 events. The Fermilab proposal P996 proposes the study of the K^+ mode at this 1000 event level

In addition to the best measurement of the Vts matrix element in the standard model, a precise measurement of . $K^+ \to\pi^+ \nu \bar{\nu}$ in the standard model is sensitive to physics beyond the standard model in loops. For example, new non-standard model particles that might be observed at the LHC might be included by adding terms as

$C_{new}X_{new}$ to the matrix element used in calculating the branching fraction above. There are many studies of for example Minimal Flavor Violation models that would be constrained by P996. A much more detailed discussion of these models is discussed in the proposal and references cited therein.

THE SIGNAL

As already discussed, the decay $K^+ \rightarrow \pi^+ \nu \bar{\nu}$ has a very small branching fraction. When trying to measure this process, other decay channels provide many sources of possible backgrounds. Figure 3 show schematically the decay modes of the K^+, including the mode of interest. In order to observe the signal, all other modes must be rejected. This is effected partly by requiring a very positive signature, and partly be rejecting unwanted modes. A positive signature is generated through the requirement that the K^+ be observed to stop in the target, and then the π^+ be observed leaving from the same region of the target followed by observing the π^+ to decay into a μ^+ and the μ^+ into an e^+ (the π-μ-e sequence). The properties of the π^+ are further constrained by measuring its momentum in the magnetic spectrometer and by its range in the range stack (see Figure 5) A trigger may be rejected if there is for example a photon signal in the veto counters at the entrance or exit of the solenoid, or if there are signals in the surrounding counters without an incoming track.

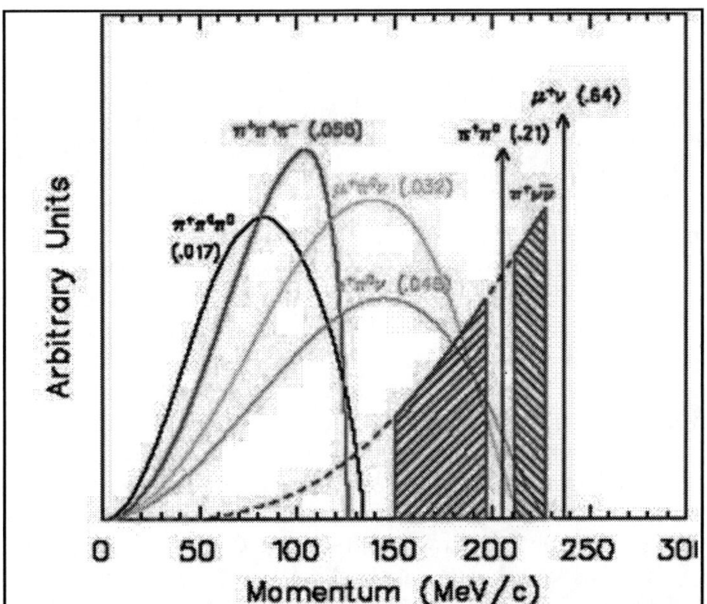

Figure 3 K+ decay modes. The vertical scale is clearly not the same for the different modes. Note that there are two possible signal regions for the K+ decay to two neutrinos. Those areas are shaded in red. The momentum scale is for the charged decay product. Numbers in parentheses

are the branching ratios. The red shaded area represents areas where the measurement can be carried out.

THE EXPERIMENT

The first and primary concern in planning an experiment is of course the rate. This concern is of course paramount when addressing a measurement where the branching fraction is so low. As noted in Figure 3, selecting signal events in this decay mode is particularly challenging.

The design of the P996 experiment follows very closely the design of the successful set of experiments carried out at Brookhaven National Laboratory - E787 and E949[5]. The discussion of the experiment P996 is a discussion of how that set of experiments can be improved upon.

There are three aspects of the experiment, each critical to the success if the experiment: the accelerator complex, the secondary K^+ beam and the detector. In each case, significant improvements are proposed. These improvements make it possible to obtain the proposed 1000 events.

Accelerator

The most dramatic improvement in the design of P996 compared to BNL-E987 is the accelerator. We have proposed to use the Fermilab Tevatron, scheduled to stop running as a collider after 2011, as a stretcher ring for the Fermilab Main Injector (MI). The MI currently provides 120 to 150 GeV proton beams to produce anti-protons and protons for the collider and proton beams to the neutrino program and a 4 second slow spill to the test beam. Typically, the MI provides a fast extracted beam to experiments every 2.3 seconds. Slow extracted beam to the test beam facility is provided once per minute. The MI will be upgraded over the next few years to decrease the cycle time so that 150 GeV protons can be delivers once per 1.667 seconds. The proposal is then to use two MI cycles to fill the Tevatron ring with 150 GeV protons (the Tevatron has twice the circumference of the MI). The beam of 96 Tp (teraprotons) would then be slowly extracted over the next 34 seconds. This scheme provides 150 GeV protons to the experiment with a duty cycle of order 95% !

Using the Tevatron ring as a stretcher ring has been tried. At the end of fixed target Tevatron running at the end of 1999 and the beginning of 2000, a test run was carried out to test the technique. 150 GeV beam was delivered to the KTeV experiment as a test run for the proposed KAMI (KAons at the Main Injector) experiment. Slow extracted beam was delivered for about one week. The test was very successful. For P996, the intensity will have to be significantly increased compared to that test run.

The K^+ Beam

The 150 GeV protons are brought to a platinum target of approximately one interaction length. The Brookhaven experiments used a 6 cm Pt target, P996 proposed a 8 cm Pt target. Because of the higher beam energy, K^+ production is improved. The overall improvement in K^+ production is expected to be improved by as much as a factor of (7 ± 2). K^+'s in the forward direction are collected, momentum selected,

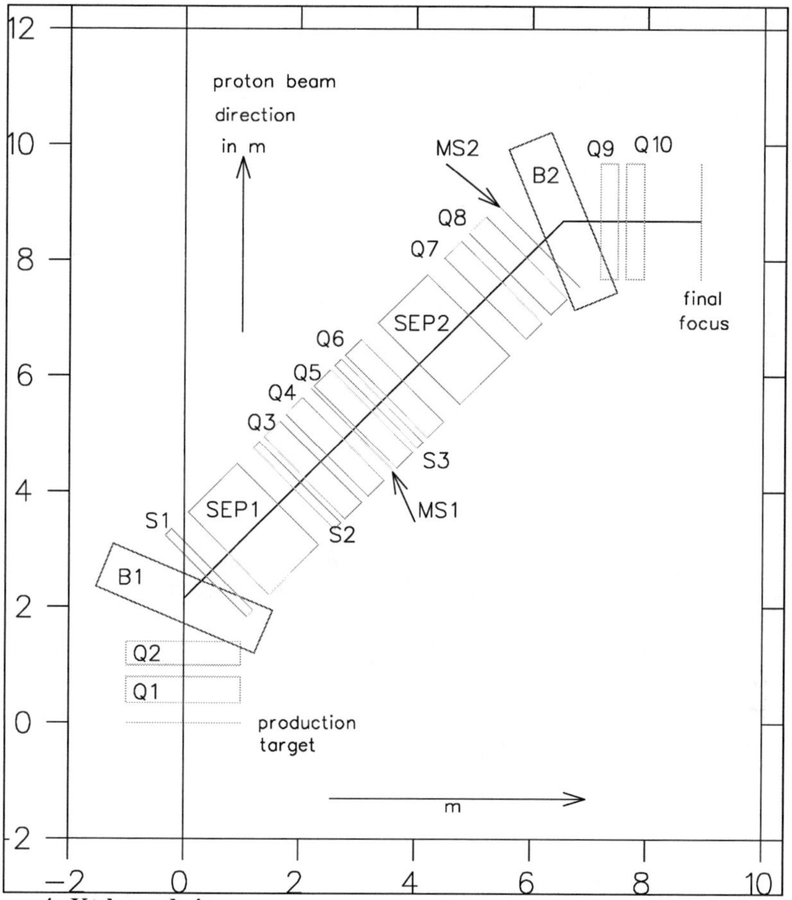

Figure 4 K+ beam design

separated, and focused onto the production target. In the BNL experiment, the secondary K+ beam was 750 MeV/c and was 19.6 meters long. Jaap Dornbos has designed a new beam[6] for this experiment. The new beam is 550 Mev/c and is 13.74 meters long. The layout of the beam is shown in Figure 4. The advantage of the lower beam momentum is that one can achieve better K/π separation, and a larger fraction of the incoming K^+ beam stops in the target. The net result is that with a somewhat higher stopping K^+ rate in the detector, the incoming flux is comparable to that in the Brookhaven experiments. As backgrounds are due primarily to the flux of incoming particles, the background is no worse than the previous experiments– even before improvements to the detector.

The Detector

The P996 detector is very similar to the detector used in the BNL experiments. The detector is shown in Figure 5. The detector is placed in a large solenoid – for example the CDF solenoid, though there are other possibilities. The K^+ beam passes through tracking, hodoscopes, and a Fitch type Čerenkov counter, and enters along the axis of the solenoid, passing through an active degrader on the way to the target where the K^+'s stop. The highly segmented active target consists of scintillator bars. The target is followed by a veto system that is used to define the stopping K+. When the K+ decays, the resulting π^+ is also detected in the target.

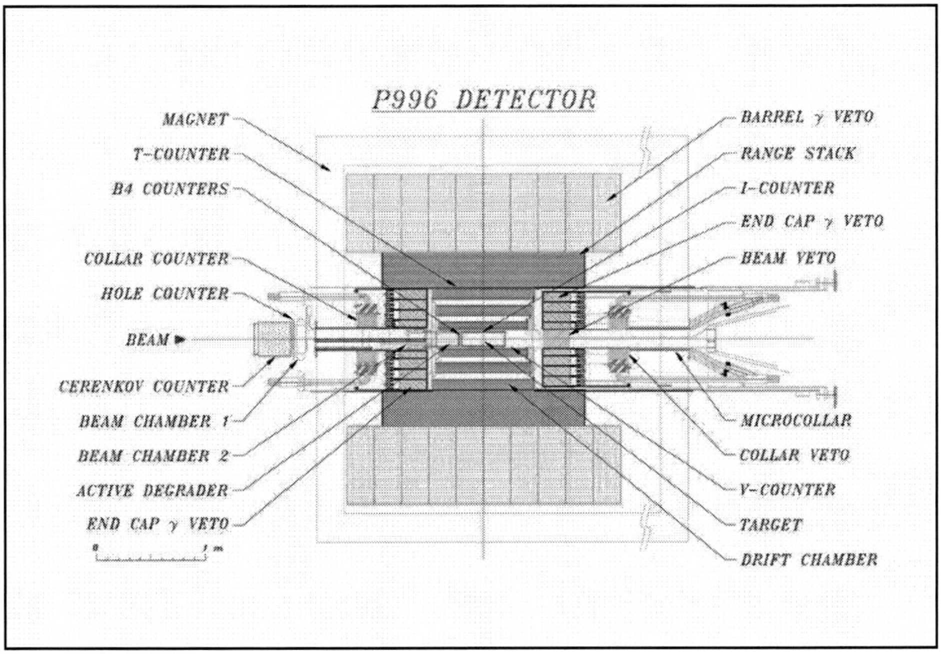

Figure 5 The P996 detector

The momentum of the π^+ is measured in the drift chamber. The π^+ stops in the range stack, which provides validation of the momentum measurement. The π^+ decay is observed within the range stack. The decay μ^+ is also observed and its range measured. The μ^+ decay to an electron is then observed.

The whole detector is surrounded by veto counters providing a full 4π coverage. The veto system is clearly critical to the suppression of background processes.

The data acquisition system will of course also be upgraded to a deadtimeless system. This will significantly improve the dead time, effectively increasing the acceptance of the detector. Timing and signal heights from all of the detectors will be saved for each event.

CONCLUSIONS

The measurement of the branching ratio of $K^+ \to \pi^+ \nu \bar{\nu}$ at the few percent level is now one of the experiments that the high energy physics community recognizes as one of the 'must be done' experiments. Proposal P996 demonstrates how this experiment might be carried out. Comparing the accelerator, beam and detector performance expected in P996 compared to BNL experiment E949, one may obtain an estimate of the $K^+ \to \pi^+ \nu \bar{\nu}$ event rate. As there have been careful calculations of the ratios of improvements in each stage, it is even possible to assign errors to the improvements and thus to the final estimate of the event rate. The number of stopping K^+ is 7.6 MHz, the running time is 5000 hours per year (approximately 60% of wall clock time), with an acceptance of about 4%, and a branching ratio of 0.85×10^{-10}, one obtains an event rate of 194^{+89}_{-79} events per year. The errors are not discussed in this document, but are discussed in detail in the full proposal. With the number of events expected in P998, the predictions based on the Standard Model will be confronted and the window will be open to the observations of new physics beyond the Standard Model.

ACKNOWLEDGMENTS

This paper is presented in memoriam of Agusto Garcia Gonzalez, a colleague who contributed in many ways to our efforts to understand the weak interaction. This talk is based on the Fermilab proposal P996 and the presentations made by R. Tschirhart, J. Ritchie and J Comfort. This work was supported by Fermilab: operated by Fermi Research Alliance, LLC under Contract No. DE-AC02-07CH11359 with the United States Department of Energy.

REFERENCES

1. L. B. Okun, *Weak Interactions of Elementary Particles*, Reading, Massachusetts; Pergamon Press 1965
2. P5 report to DOE at http://www.er.doe.gov/hep/panels/reports/hepap_reports.shtml
3. F. Mescia and C. Smith, Phys. Rev **D 76**, 034017 (2007) [arXiv:0705.2025 [hep-ph]], and other papers referenced in the P996 proposal
4. Joseph Comfort et. al Measurement of $K^+ \to \pi^+ \nu \bar{\nu}$ Decay at Fermilab, Fermilab Proposal P996, October 14, 2009 see also references cited therein
5. A. V. Artamonov et al., *Phys Rev* **79**, 092004 (2009).
6. Jaap Dornboss "A 550 MeV/c kaon beam" TRIUMPH, September 16, 2009, unpublished

Accelerators Beyond The Tevatron?

Joseph Lach

Fermi National Accelerator Laboratory
P.O Box 500, Batavia Illinois

Abstract. Following the successful operation of the Fermilab superconducting accelerator three new higher energy accelerators were planned. They were the UNK in the Soviet Union, the LHC in Europe, and the SSC in the United States. All were expected to start producing physics about 1995. They did not. Why?

Keywords: UNK; SSC;LHC; Tevatron.
PACS: 01.10.Fv

JOURNEY: 6.2 BEV →33 GEV → 200 → 400 → TEV

In the spring of 1963 I completed my thesis work at the Lawrence Radiation Laboratory (LRL) in Berkeley. It used the 6.2 BeV accelerator – The Bevatron - then the world's highest energy proton accelerator. I then joined the Yale University Physics Department and was a builder of one of the first rf-separated beams[1] at the Alternation Gradient Synchrotron (AGS) at Brookhaven National Laboratory - then the highest energy (33 GeV) proton accelerator.

I joined the 1965 "summer study" at LRL to work on experimental beams for the LRL design of the proposed 200 BeV Accelerator. At the end of that summer, the LRL director, received a letter from an unknown – at least to me – physicist from a small university in upstate New York. In his scathing letter[2] Robert Wilson said their accelerator design was "much too conservative" and "lacking in imagination".

In the summer of 1969 I joined the National Accelerator Laboratory (NAL) in Batavia IL. Robert Wilson, now Director, had decided that the 200 BeV Accelerator would now be a 400 BeV accelerator. The construction schedule would be advanced by one year[2] and this would be done with a reduced construction cost!

The NAL accelerator achieved 200 GeV beam in March 1972 and a full scale research program at 400 GeV by 1975.

The NAL overview in Figure 1 shows the four mile circumference ring and experimental areas. Figure 2 shows the interior of the accelerator tunnel and the conventional magnets used for the program just described. But now one also sees the second lower ring of superconducting magnets that evolved into the Tevatron.

CP1259, *First CINVESTAV-UNAM Symposium on High Energy Physics*
edited by J. C. D'Olivo, R. López-Fernández, and M. Á. Pérez

Figure 1. The four mile ring contains the conventional 400 GeV accelerator.

Figure 2. Superconducting magnets below the conventional magnets.

THREE NEW ACCELERATORS

In 1986 Fermilab's high-energy superconducting proton accelerator (Tevatron) began operation at 400 GeV. In 1987 operation began at 800 GeV. This successful operation of a large superconducting accelerator energized the high-energy physics community into a new era of accelerator construction. The recorder of progress in this field, the Particle Data Group at LBL published the following table in their 1988 edition.

HIGH-ENERGY COLLIDER PARAMETERS (Cont'd)

***pp*, *$\bar{p}p$*, and *ep* Colliders**

The numbers here were received from representatives of each collider by late 1987. Numbers are subject to change, and many are only estimates. (Quantities are, where appropriate, r.m.s. *H* ≡ horizontal direction, *V* ≡ vertical direction, s.c. ≡ superconducting.)

	$Sp\bar{p}S$ (CERN)	TEVATRON (Fermilab)	HERA (DESY)	UNK (Serpukhov)	LHC (CERN)		SSC (USA)
Physics start date	1981	1987	Spring 1990	1995 ?	1995 ?		1996 ?
Particles collided	$p\bar{p}$	$p\bar{p}$	*ep*	*pp*	*pp*	*ep*	*pp*
Max. beam energy (TeV)	0.315 (0.45 in pulsed mode)	0.9–1.0	*e*: 0.026 *p*: 0.82	3	8	*e*: 0.05 *p*: 8	20
Injection energy (TeV)	0.026	0.15	*e*: 0.014 *p*: 0.040	0.4	0.450	*e*: 0.02 *p*: 0.450	1
Luminosity (10^{30} cm^{-2} s^{-1})	0.3 3 (1988)	0.5–1.0	15	400	1400	200	1000, β^* = 0.5 m 56, β^* = 10 m
Circumference (km)	6.911	6.28	6.336	20.772	26.659		83.631
No. of interaction regions	2	2 high $\mathcal{L}$ 2 low $\mathcal{L}$	3	4	7	3	4 (initially)
No. of particles per bunch (units 10^{10})	*p*: 15 $\bar{p}$: 2–10	*p*: 6 $\bar{p}$: 2	*e*: 3.48 *p*: 10	3	2.5	*e*: 8 *p*: 30	0.80
No. of bunches per ring per species	3–6	3–6	220	8,000	3564	540	15,456

Figure 3. High-Energy Accelerator parameters 1988

Note that three new accelerators appear in the above figure. They are the UNK being constructed at Serpukov in the Soviet Union, the Large Hadron Collider (LHC) in Europe at CERN, and the Superconducting Super Collider (SSC) at a not yet chosen U.S. site. The physics start date for each of these accelerators is estimated to be 1995-6. These are each major scientific projects with costs in billions of dollars.

The following table contains a time line for the new accelerator construction and some other relevant events.

Accelerator Time Lines		
Jul-83	TM-1218 Tunneling Beyond Fermilab Site	Joe Lach
Jan-84	Construction of UNK starts, 20.8 km ring	Victor Yarba
Jun-84	SSC Central Design Group at LBL started	Maury Tigner
Nov-84	Ronald Reagan reelected President	
Jan-87	SSC site selection, 43 proposals 87.1 km ring	
Nov-88	George H. Bush elected President, Texas SSC site chosen	
Aug-89	LEP starts operation, 26.7 km ring	
Sep-89	Berlin wall falls	
Dec-91	Soviet Union disintegrates	
Jan-92	UNK construction ends	Victor Yarba
Nov-92	W. Clinton, A. Gore elected President and Vice-President	
Sept-93	US/Russia agree to build Space Station	Gore/Chernomyrdin
Nov-93	Texas SSC terminated	
Jun-99	First Space Station module, Zarya launched	
Dec-00	LEP Shuts down	
Sept-08	LHC starts, quench incident, LHC down	
Nov-09	LHC resumes operation??	

Table 1. Accelerator Time Lines

UNK

The UNK has a ring about three times larger than the Fermilab accelerator. Like the Tevatron UNK will contain a ring of conventional magnets to bring protons to 400 GeV and a superconducting ring which will accelerater protons to 3 TeV. Thus in many ways it will be a Tevatron scaled up by a factor of three. In 1985 I visited the construction site and the imposing tunnel shown in Figure 4. Figure 5a: Gennady Gurov stands before a conventional dipole; Figure 5b: the coil of a superconducting dipole.

I was impressed with the UNK project; the tunnel was steel lined. It was of generous proportions – I suspect that Bob Wilson would have objected to that – but servicing and upgrades would have been much easier. I recall commenting that they should keep to their schedule. They would only be able to keep the title of “highest energy’ for a short time since Fermilab would build the SSC with an aggressive construction schedule – like Bob Wilson!

The disintegration of the USSR in late 1991, with the ensuing economic crisis, forced the closing of the UNK project. Most of the tunnel and about 80% of the conventional accelerator had been installed, and a 25 magnet superconducting string had been successfully tested.

Figure 4, The UNK tunnel under construction

Figure 5a UNK conventional magnet. Figure 5b superconducting coil

SSC

We had all heard about the SSC. We had been to workshops, summer studies, etc. helping to fix the parameters of such a machine. We were even more excited about the great physics that it would produce. Where would it be sited? Who would build it? Might it be at Fermilab? Those were the really big questions.

In the spring of 1983, Leon Lederman, then director of Fermilab, asked me to form a small group and address the question of whether Fermilab might be an appropriate site for the SSC? We knew it would be a 20 TeV accelerator with a circumference of about 52 miles. It certainly could not be contained within the Fermilab 6800 acre site. Could we go beyond the present site? What would the tunnel cost? Would it be on the surface like the Tevatron or a deeper tunnel? What was the geology like? Address the radiation safety problems, cost, location, etc.

The result was short paper that is the first entry in Table 1 and is reference 3. It is that document that led – about five years later - to the Illinois SSC proposal.

Within 50 miles of Fermilab is Chicago's Tunnel and Reservoir Plan (TARP). It is a water reclamation project that utilizes tunnels in dolomite rock under the Chicago region. Some of the almost 100 miles of tunnel are over 35 feet in diameter. This same rock stratum extends under Fermilab and the area of the proposed Illinois SSC site.

A major step, shown in Table 1, was the establishment of a Central Design Group at LBL headed by Maury Tigner. This led to a specific SSC design. I recommend[4] the excellent history of the Central Design Group by Stanley Wojcicki.

Continuing in Table 1 we see that in 1984 Ronald Reagan was elected to a second term and in his administration, the SSC site selection process was initiated. A total of 43 proposed sites were considered; a major national event patterned after the site selection process for Fermilab in 1967. A group, SSC For Illinois, Inc. was constituted to write the proposal. It would have been naïve to think that perception and politics would not be involved in the selection process. I recall a meeting where someone suggested that the major highway bordering Fermilab be named the "Ronald Reagan Memorial Highway" - because some other states that were making proposals were making similar changes. I-88 is still called the Ronald Reagan Memorial Highway!

In hindsight it should not have been a great surprise that Texas was chosen as the SSC site on the day that George H. Bush was elected president.

The new SSC site in Waxahachie Texas had many problems. There were organizational issues, major cost overruns, and stretched out schedules. These are described in a second article[5] by Stanley Wojcicki.

In 1992 W. Clinton was elected president. In 1993 the US and Russia signed an agreement to build the Space Station. This major and costly project of the new administration and the discouraging progress of the SSC prompted Congress to terminate funding for the SSC in 1993.

Stanley Wojcicki[5]: " **In retrospect, choosing a green site in Texas was a mistake**".

This mistake cost billions of dollars and set back high energy physics by at least a decade. **We need to ask why, who, and how this mistake was made.**

LHC

In 1986 CERN was constructing LEP; the world's largest accelerator for the collisions of electrons and positrons. Its 26.7 km length would extend into France well beyond the borders of CERN's existing Swiss site. This site would be pushing the limits of such a ring since it would extend into the challenging geology of the Jura mountains. Figure 6 taken during the construction in 1986 shows water inflow that delayed construction but was eventually contained.

The successful LEP program (Table 1) began in 1989 and concluded at the end of 2000. An important decision was made to use the LEP tunnel for the LHC. That entailed a removal of the smaller LEP magnets and the installation of the much larger LHC magnets and the construction of the two large caverns to house the massive CMS and ATLAS detectors. This enormous project took about seven years to complete.

Figure 6. LEP tunnel water in the Jura region.

This new configuration for the LHC is shown in Figure 7. The proposed SSC ring size is a factor of 3.25 larger than the new LHC ring so that even when the LHC is operating at full energy, 7.0 TeV, it would be less than 1/3 of the SSC energy. To reach 7.0 TeV, the LHC magnets must operate at 8.3 T. The late SSC would operate at only 6.79 T at full energy. Thus the LHC requires a truly "heroic" cryogenic system. Figure 8, taken from the CERN web page "Facts and Figures" describes the magnitude of the cryogenic system.

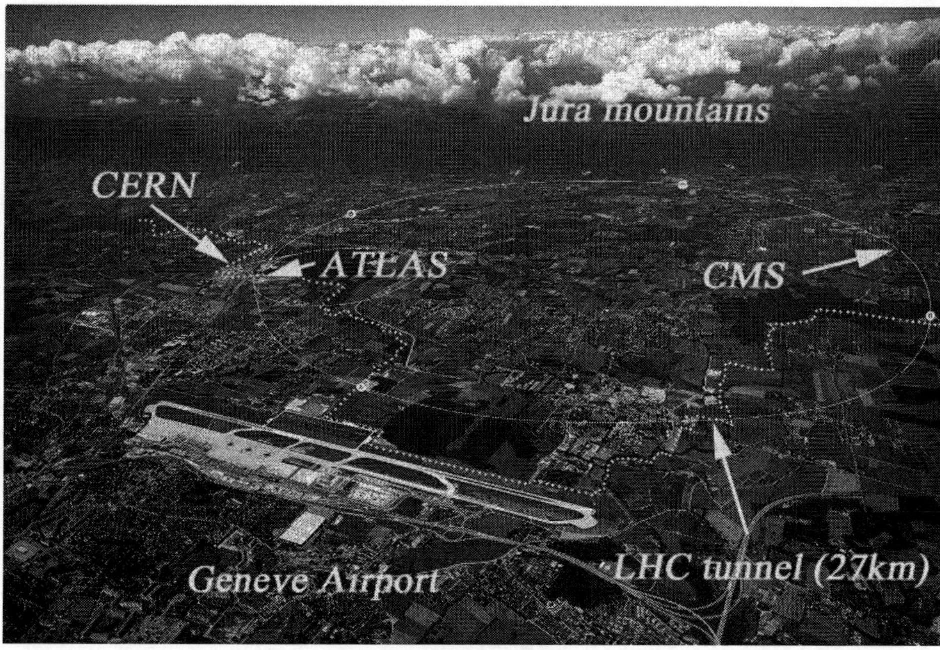

Figure 7. The LHC tunnel, detectors and the Jura.

Facts and figures

The largest machine in the world...

The precise circumference of the LHC accelerator is 26 659 m, with a total of 9300 magnets inside. Not only is the LHC the world's largest particle accelerator, just one-eighth of its cryogenic distribution system would qualify as the world's largest fridge. All the magnets will be pre-cooled to -193.2°C (80 K) using 10 080 tonnes of liquid nitrogen, before they are filled with nearly 60 tonnes of liquid helium to bring them down to -271.3°C (1.9 K).

Figure 8. **LHC cooling**

The LHC started operation in the September 2008 and suffered a major quench incident that damaged a couple dozen magnets and vented several tons of He. The LHC does not fail gracefully!

The recovery in late 2009 – at half energy – is proceeding cautiously. We hope and expect that the LHC will recover and become a robust, healthy and productive machine. With the loss of the SSC and UNK, the LHC remains our only hope for an accelerator beyond the Tevatron.

ACKNOWLEDGMENTS

I am indebted to Victor Yarba for information about UNK; the photos are my own. Comments from Alexei Vorobyov on the history of UNK were appreciated.

Many people shared their thoughts on the SSC including John Peoples, Chris Quigg and Bruce Chrisman. I thank Stan Wojiciki for his insightful history articles.

I thank Chris Laughton for use of his LEP tunnel picture. Other CERN pictures and words came from the CERN web site.

I thank Miguel A. Perez for inviting me to this meeting and pressing me to record my presentation. This is a fitting way to honor the memory of our friend Augusto Garcia.

I also wish to thank Bob Wilson for teaching us how to build accelerators with beauty and economy.

REFERENCES

1. H. W. J. Foelsche, H. Hahn, H. J. Halama, J. Lach. T. Ludlam, and J. Sandweiss, **Radio Frequency Separated Beam at the AGS**, Rev. Sci. Instr. 38, 879 (1967)
2. Lillian Hoddeson, Adrienne W. Kolb, and Catherine Westfall, Fermilab Physics, The Frontier and Megascience. The University of Chicago Press 2008.
3. S. Baker, A. Elwyn, J. Lach, A. Read, **Tunneling Beyond The Fermilab Site.** Fermilab TM-1218, July 1, 1983.
4. Stanley Wojeicki, **The Supercollider: The Pre-Texas Days**. Reviews of Accelerator Science and Technology, Vol. 1 (2008) 259-302
5. Stanley Wojeicki, **The Supercollider: The Texas Days**. Reviews of Accelerator Science and Technology, Vol. 2 (2009) 1-37

PHENOMENOLOGY

Intermediate p_T phenomena in relativistic heavy-ion collisions[1]

Alejandro Ayala

Instituto de Ciencias Nucleares, Universidad Nacional Autónoma de México, Apartado Postal 70-543, México Distrito Federal 04510, Mexico.

Abstract. Particles produced with intermediate values of transverse momentum $1-2 \lesssim p_T \lesssim 4-5$ GeV, exhibit both thermal and jet-like features making them an ideal testing ground of the properties of the quark-gluon plasma produced in the aftermath of a relativistic heavy-ion collision, which have been only inferred from the behavior of either soft or hard modes. In this work I discuss two phenomena that that are observed in this transverse momentum window: the suppression of the single, non-photonic electron yield compared to normalized p+p collisions and the spin alignment of vector mesons. I make use of a recombination scenario of either a heavy or a fast quark with thermal ones and show that this way of describing hadron formation can account for the afore mentioned phenomena.

Keywords: Recombination, nuclear modification factor, spin alignment
PACS: 25.75.-q, 13.85.Qk, 25.75.Nq, 25.75.Ld

INTRODUCTION

Many interesting phenomena have been observed in high energy heavy-ion collisions at the Relativistic Heavy-Ion Collider (RHIC). The suppression of the single hadron transverse momentum spectra in Au + Au collisions and the anomalous suppression of single, non-photonic electrons, as compared to normalized p + p collisions, are two of such phenomena. The usual interpretation of these results is given in terms of the energy loss of fast partons traversing the medium and multiply scattering (both elasticly and inelasticly) from medium constituents [1, 2]. These phenomena are clearly observed at what one can properly dub the *high* p_T region, which consists of particles with final momenta $p_T \gtrsim 4-5$ GeV. On the other hand in the *low* p_T region, where final particle momenta are such that $p_T \lesssim 1-2$ GeV there is a host of phenomena described in terms of collective behavior such as transverse and elliptic flow, associated with the thermal properties of the matter formed in the aftermath of the collision [3, 4].

Nevertheless, the properties of the final state hadrons whose spectra lies in the *intermediate* p_T region, $1-2 \lesssim p_T \lesssim 4-5$ GeV, are less well understood in terms of either hard or soft processes. Moreover, this is the region where the recombination scenario [5] merges with the fragmentation one for particle production. It is only natural that particles in this momentum window exhibit both collective as well as jet-like features, given that their constituent quarks partake of these kind of processes before they find themselves

[1] In memoriam of Augusto García whose conversations helped me to understand bits of physics but more importantly, how science is the product not only of the mind but specially of the soul.

CP1259, *First CINVESTAV-UNAM Symposium on High Energy Physics*
edited by J. C. D'Olivo, R. López-Fernández, and M. Á. Pérez

making up the final hadrons.

In this contribution, I examine two scenarios that address particle formation in the intermediate p_T region. These make use of the recombination mechanism between a hard (or heavy) –that is, not thermalized– quark and soft ones. The first scenario describes the effect of the enhancement of the baryon to meson ratio to explain the suppression of the single, non-photonic electron yield. The second one explains the polarization properties of vector mesons by invoking the Thomas spin precession during hadronization. At the end, I present an outlook intended to call attention to the issue of how these kind of observables can help disentangling the physical mechanisms responsible for particle production in the intermediate p_T region.

DYNAMICAL HEAVY-QUARK RECOMBINATION AND THE SINGLE, NON-PHOTONIC ELECTRON NUCLEAR MODIFICATION FACTOR

The suppression of single, non-photonic electrons at RHIC [6, 7] is usually attributed to heavy-quark energy losses. However, calculations that successfully describe the nuclear modification factor of hadrons fail to describe R^e_{AA} for the single, non-photonic electrons [8, 9, 10]. This has prompted a great deal of effort aimed to better describe the heavy-quark energy loss mechanisms to include not only the radiative part [11, 12, 13, 14] but also the collisional [15, 16] and the medium dynamical properties [17] to compute the radiative piece. As a result, although some improvement in the description of the nuclear modification factor has been gained, it is not yet clear whether the anomalous suppression can be completely attributed to energy losses.

Working along a complementary approach to describe the non-photonic electron yield at RHIC, it has been argued [18, 19] that under the assumption of an enhancement in the heavy-quark baryon to meson ratio, analogous to the case of the proton to pion and the Λ to kaon ratios in Au + Au collisions [20, 21, 22, 23], it is possible to achieve a larger suppression of R^e_{AA}. The rationale behind the idea is that heavy-quark mesons have a larger branching ratio to decay inclusively into electrons as compared to heavy-quark baryons, and therefore, when the former are less copiously produced in a heavy-ion environment, the nuclear modification factor decreases, even in the absence of heavy quark energy losses in the plasma.

In order to give a qualitative argument that shows how an enhancement in the heavy-quark baryon to meson ratio can suppress the single, non-photonic electron nuclear modification factor, let us look at the p_T integrated R^e_{AA} and to consider that the heavy hadrons are only those containing a single charm,

$$R_{AA}^{e\ p_{T\,\mathrm{int}}} = \frac{1}{\langle n_p \rangle} \frac{N_{AA}^{\Lambda} B^{\Lambda \to e} + N_{AA}^{D} B^{D \to e}}{N_{pp}^{\Lambda} B^{\Lambda \to e} + N_{pp}^{D} B^{D \to e}}, \tag{1}$$

where $\langle n_p \rangle$ is the average number of participants in the collision for a given centrality class, $N^x_{AA\ (pp)}$, refers to the number of x-particles produced in A + A (p + p) collisions and $B^{x \to e}$ is the branching ratio for the inclusive decay of x-particles into electrons. Let us bring Eq. (1) into a form that contains the corresponding p_T integrated nuclear modification factor for particles containing charm. We write

$$R_{AA}^{e\ p_{T\,\mathrm{int}}} = \frac{1}{\langle n_p \rangle} \left(\frac{N_{AA}^D}{N_{pp}^D} \right) \left(\frac{B^{D\to e} + \frac{N_{AA}^{\Lambda}}{N_{AA}^{D}} B^{\Lambda\to e}}{B^{D\to e} + \frac{N_{pp}^{\Lambda}}{N_{pp}^{D}} B^{\Lambda\to e}} \right). \tag{2}$$

Let us introduce the shorthand notation

$$\begin{aligned} C &= \frac{N_{AA}^{\Lambda}/N_{AA}^{D}}{N_{pp}^{\Lambda}/N_{pp}^{D}} \\ x &= \frac{B^{\Lambda\to e}}{B^{D\to e}}, \end{aligned} \tag{3}$$

where C represents the *enhancement factor* for the ratio of charm baryons to mesons in A + A as compared to p + p collisions and x is the charm baryon to meson relative branching ratios for their corresponding inclusive decays into electrons. With these definitions, and after rewriting the factor N_{AA}^D/N_{pp}^D in the form

$$\begin{aligned} \frac{N_{AA}^D}{N_{pp}^D} &= \frac{N_{AA}^D + N_{AA}^{\Lambda} - N_{AA}^{\Lambda}}{N_{pp}^D + N_{pp}^{\Lambda} - N_{pp}^{\Lambda}} \\ &= \left(\frac{N_{AA}^D + N_{AA}^{\Lambda}}{N_{pp}^D + N_{pp}^{\Lambda}} \right) \left(\frac{1 - N_{AA}^{\Lambda}/(N_{AA}^D + N_{AA}^{\Lambda})}{1 - N_{pp}^{\Lambda}/(N_{pp}^D + N_{pp}^{\Lambda})} \right), \end{aligned} \tag{4}$$

we can express Eq. (2) as

$$\begin{aligned} R_{AA}^{e\ p_{T\,\mathrm{int}}} &= \frac{1}{\langle n_p \rangle} \left(\frac{N_{AA}^D + N_{AA}^{\Lambda}}{N_{pp}^D + N_{pp}^{\Lambda}} \right) \left(\frac{1 - N_{AA}^{\Lambda}/(N_{AA}^D + N_{AA}^{\Lambda})}{1 - N_{pp}^{\Lambda}/(N_{pp}^D + N_{pp}^{\Lambda})} \right) \left(\frac{1 + CxN_{pp}^{\Lambda}/N_{pp}^D}{1 + xN_{pp}^{\Lambda}/N_{pp}^D} \right) \\ &\equiv \frac{1}{\langle n_p \rangle} \left(\frac{N_{AA}^D + N_{AA}^{\Lambda}}{N_{pp}^D + N_{pp}^{\Lambda}} \right) T_{AA}^{e\ p_{T\,\mathrm{int}}}. \end{aligned} \tag{5}$$

When not integrated over transverse momentum, the factor

$$R_{AA}^{c} = 1/\langle n_p \rangle \left[(N_{AA}^D + N_{AA}^{\Lambda})/(N_{pp}^D + N_{pp}^{\Lambda}) \right] \tag{6}$$

represents the nuclear modification factor for particles with charm. Let us not assume any particular value for this factor and instead concentrate in the other one in Eq. (5), which can be written as

$$T_{AA}^{e\ p_{T\,\mathrm{int}}} = \frac{(1+a)(1+xCa)}{(1+Ca)(1+xa)}, \tag{7}$$

where $a = N_{pp}^{\Lambda}/N_{pp}^{D}$. It is easy to see that the function $T_{AA}^{e\ p_{T\,\mathrm{int}}}$ is less than 1 when $x < 1$ provided that $Ca > a$.

To quantitatively address the question of whether the enhancement factor C times a –namely, the heavy-baryon to heavy-meson ratio in Au + Au collisions– can indeed be larger than a –namely, the heavy-baryon to heavy-meson ratio in p + p collisions– and if so, how this affects the behavior of the factor T^e_{AA} as a function of p_T, we use a dynamical recombination scenario [24] that accounts for the fact that the probability to form baryons and mesons can depend on a different way on the evolving density during the collision. A coalescence model addressing the same goals has been recently presented in Ref. [25].

The upshot of the DQRM is that the density evolving probability differs for hadrons made up by two and three constituents *with the same mass*, that is to say, the relative population of baryons and mesons can be attributed not only to flow but rather to the dynamical properties of quark clustering in a varying density scenario. A natural question is whether those features remain true for baryons and mesons with one constituent heavy-quark and whether a computed, as opposed to assumed, baryon to meson ratio, can at least partially explain the anomalous single, non-photonic electron suppression at RHIC.

Recall that for the kinematic regime where quark recombination happens, the assumption is that the recombining light flavors are thermal. On the other hand, the thermalization of heavy flavors has been a subject of intense research over the past years [26, 27]. Nevertheless, it is not clear whether heavy flavors need to be thermal to recombine or else, whether recombination can happen in a sort of *pick up* reaction where a heavy flavor, not necessarily thermal, coalesces with light quarks on its way out to form either mesons or baryons. Here we employ the point of view that the recombination of heavy quarks happens without the need of these to be initially thermal, though the light quarks that the heavy quark coalesces with have already thermalized, producing a finally thermal hadron.

The invariant transverse momentum distribution of a given hadron can be written as an integral over the freeze-out space-time hypersurface Σ, of the relativistically invariant phase space particle density $F(x,P)$,

$$E\frac{dN}{d^3P} = g\int_{\Sigma_f} d\Sigma\, \frac{P\cdot u(x)}{(2\pi)^3} F(x,P)\,, \tag{8}$$

where P is the hadron's momentum, $u(x)$ is a future oriented unit four-vector normal to Σ and g is the degeneracy factor for the hadron which takes care of the spin degree of freedom. The function $F(x,P)$ contains the information on the probability that the given hadron is formed.

To allow for a dynamical recombination scenario in a thermal environment, let us assume that the phase space particle density $F(x,P)$ can be factorized into the product of a term containing the thermal occupation number, including the effects of a possible flow velocity, and another term containing the system energy density ε driven probability $\mathscr{P}(\varepsilon)$, for the coalescence of partons into a given hadron. We thus write

$$F(x,P) = e^{-P\cdot v(x)/T}\,\mathscr{P}(\varepsilon)\,, \tag{9}$$

where $v(x)$ is the flow velocity. The probability $\mathscr{P}(\varepsilon)$ incorporates in a simple manner the information that the coalescing partons need to be close in configuration space as well as to have a not so different velocity.

To compute the probability $\mathscr{P}(\varepsilon)$, it has been shown in Ref. [24, 28] (where we refer the reader to for details) that use can be made of the *string flip model* [29, 30, 31] in order to get information about the likelihood of clustering of constituent quarks to form hadrons from an effective quark-quark interaction. In short, the model is a variational quantum Monte Carlo simulation that, taking a set of equal number of all color quarks and antiquarks at a given density, computes the optimal configuration of colorless clusters (baryons or mesons) by minimizing the potential energy of the system. At low densities, the model describes the system of quarks as isolated hadrons while at high densities, this system becomes a free Fermi gas of quarks. The interaction between quarks is pair-wise and taken as harmonic. The optimal clustering is achieved by finding the pairing producing the minimum in the potential energy between two given sets of quarks of different color for all possible color charges.

All the results we present here come from simulations done with 384 particles, 192 quarks and 192 antiquarks, corresponding to having 32 light quarks and 32 heavy quarks, plus their antiquarks in the three color charges (anti-charges). Hereafter we refer to light quarks as u-quarks and to the heavy ones as c-quarks. The number of quarks corresponds to the second closed shell of a three-dimensional box. We have checked that for this number of particles, the errors associated with finite size are already negligible. The equal amounts of u and c quarks used in the simulation is not intended to represent the whole system but rather the relative fraction which drives the recombination. In the simulation the relative abundance of baryons and mesons containing one c-quark depends on the initial number of u and c-quarks. We choose to work with equal numbers of each kind of quarks and at the end convert the relative fraction of resulting baryons and mesons to implement the physical conditions requiring that the number of c-quarks be small compared to the number of u-quarks. To take into account the mass difference between the u and c quarks we set $m_c = 10m_u$. We have checked that variations of this particular choice do not affect our relative probabilities.

In order to quantify how the different probabilities to produce sets of three quarks as compared to sets of two quarks affect the particle's yields as the energy density changes during hadronization, we need to resort to a model for the space-time evolution of the collision. We take Bjorken's scenario which incorporates the fact that initially, expansion is longitudinal, that is, along the beam direction which we take as the $\hat{z}$ axis and include transverse flow as a small effect on top of the longitudinal expansion. In this scenario, the relation between the temperature T and the 1+1 proper-time τ is given by

$$T = T_0 \left(\frac{\tau_0}{\tau}\right)^{v_s^2}, \tag{10}$$

where $\tau = \sqrt{t^2 - z^2}$. Equation (10) assumes that the speed of sound v_s changes slowly with temperature. For simplicity we take v_s as a constant equal to the ideal gas limit $v_s^2 = 1/3$.

We also consider that hadronization takes place on hypersurfaces Σ characterized by a constant value of τ and therefore

$$d\Sigma = \tau\rho \, d\rho \, d\phi \, d\eta, \tag{11}$$

where

$$\eta = \frac{1}{2}\ln\left(\frac{t+z}{t-z}\right), \tag{12}$$

is the spatial rapidity and ρ, ϕ are the polar transverse coordinates. Thus, the transverse spectrum for a hadron species H is given as the average over the hadronization interval of the right hand-side of Eq. (8), namely

$$E\frac{dN^H}{d^3P} = \frac{g}{\Delta\tau}\int_{\tau_0}^{\tau_f} d\tau \int_\Sigma d\Sigma \, \frac{P\cdot u(x)}{(2\pi)^3} F^H(x,P), \tag{13}$$

where $\Delta\tau = \tau_f - \tau_0$.

To find the relation between the energy density ε –that the probability $\mathscr{P}$ depends upon– and T, we resort to lattice simulations. For the case of two flavors (since the heavy quark does not thermalize), a fair representation of the data [32] is given by the analytic expression

$$\varepsilon/T^4 = a\left[1+\tanh\left(\frac{T-T_c}{bT_c}\right)\right], \tag{14}$$

with $a = 4.82$ and $b = 0.132$. We take $T_c = 175$ MeV.

The flow four-velocity vector v^μ is given by

$$v^\mu = (\cosh\eta\cosh\eta_T, \sinh\eta_T\cos\phi, \sinh\eta_T\sin\phi, \sinh\eta\cosh\eta_T), \tag{15}$$

where the magnitude of the transverse flow velocity v_T and η_T are related by $v_T = \tanh\eta_T$. The normal to the freeze-out hypersurfaces of constant τ, u^μ, is given by

$$u^\mu = (\cosh\eta, 0, 0, \sinh\eta). \tag{16}$$

We write the momentum four-vector in components as

$$P^\mu = (m_T\cosh y, p_T\cos\Phi, p_T\sin\Phi, m_T\sinh y), \tag{17}$$

where y is the $1+1$ momentum rapidity given by

$$y = \frac{1}{2}\ln\left(\frac{E+p_z}{E-p_z}\right), \tag{18}$$

and Φ the azimuthal angle of the momentum components in the transverse plane.

Therefore, the products $P\cdot u$ and $P\cdot v$ appearing in Eq. (13) can be written as

$$\begin{aligned} P\cdot v &= m_T\cosh(\eta-y)\cos\eta_T - p_T\cos(\phi-\Phi)\sinh\eta_T \\ P\cdot u &= m_T\cosh(\eta-y), \end{aligned} \tag{19}$$

Considering the situation of central collisions, we can assume that there is no dependence of the particle yield on the transverse polar coordinates. Integration over these variables gives

$$\frac{dN}{p_T dp_T dy} = g\frac{m_T}{4\pi}\frac{\rho^2_{\text{nucl}}}{\Delta\tau}\int_{\tau_0}^{\tau_f}\tau d\tau \mathscr{P}(\tau) I_0(p_T \sinh\eta_T/T)$$
$$\times \int d\eta \cosh(y-\eta) e^{-[m_T\cosh(y-\eta)\cosh\eta_T]/T} \tag{20}$$

where ρ_{nucl} is the radius of the colliding nuclei and I_0 is the Bessel function I of order zero.

We now consider as a further simplification that the space-time and momentum rapidities are completely correlated, that is $\eta \sim y$. Under this assumption, the integral over η in Eq. (20) can be performed and we finally get

$$\frac{dN}{p_T dp_T dy} = g\frac{m_T\Delta y}{4\pi}\frac{\rho^2_{\text{nucl}}}{\Delta\tau}\int_{\tau_0}^{\tau_f}\tau d\tau \mathscr{P}(\tau) I_0(p_T \sinh\eta_T/T) e^{-m_T\cosh\eta_T/T}. \tag{21}$$

Armed with the expression to compute the hadron transverse momentum distribution, we now proceed to apply the analysis to the computation of the charmed meson and baryon distributions.

We set the masses of the charmed baryons and mesons as $m^B = 2.29$ GeV (corresponding to Λ_c) and $m^M = 1.87$ GeV (corresponding to D). We take the initial hadronization time as $\tau_0 = 1$ fm, at an initial temperature $T_0 = 200$ MeV and the final hadronization temperature as $T_f = 100$ MeV, corresponding, according to Eq. (10), to a final time $\tau_f = 8$ fm.

Figure 1 shows the charmed baryon to meson ratio obtained from the ratio of the above transverse momentum distributions. Shown is a range for this ratio when varying the transverse expansion velocity v_T from 0 to 0.4. Notice that for a finite v_T, this ratio goes above 1 for $p_T \gtrsim 3.5$ GeV.

We now proceed to compute the p_T unintegrated function T^e_{AA}. For this purpose, we take that the possible charmed mesons decaying inclusively into electrons or positrons are $D^\pm$ ($B^{D^\pm\to e^\pm} = 16.0\%$), $D^0, \bar{D}^0$ ($B^{D^0,\,\bar{D}^0\to e^\pm} = 6.53\%$), $D_s^\pm$ ($B^{D_s^\pm\to e^\pm} = 8\%$) and that the possible charmed baryons decaying inclusively into electrons or positrons are $\Lambda_c, \bar{\Lambda}_c$ ($B^{\Lambda_c,\,\bar{\Lambda}_c\to e^\pm} = 4.5\%$). Thus we get

$$x = 0.14. \tag{22}$$

We also approximate the masses of all the charmed mesons considered to be equal to the mass of the $D^\pm$ mesons.

From Eq. (7) we see that, without integrating over p_T, the dependence on the transverse momentum comes from $a = (dN^\Lambda_{pp}/dp_T)/(dN^D_{pp}/dp_T)$ and the product $Ca = (dN^\Lambda_{AA}/dp_T)/(dN^D_{AA}/dp_T)$. The integrated ratio a^{int} has been computed in Ref. [19] using a Pythia simulation, with the result $a^{\text{int}} = 0.073$. We have also performed a simulation using Pythia at NLO with 100,000 events and have found that with such statistics, the ratio of charmed baryons to charmed mesons in p + p collisions at $\sqrt{s_{NN}} = 200$ GeV is

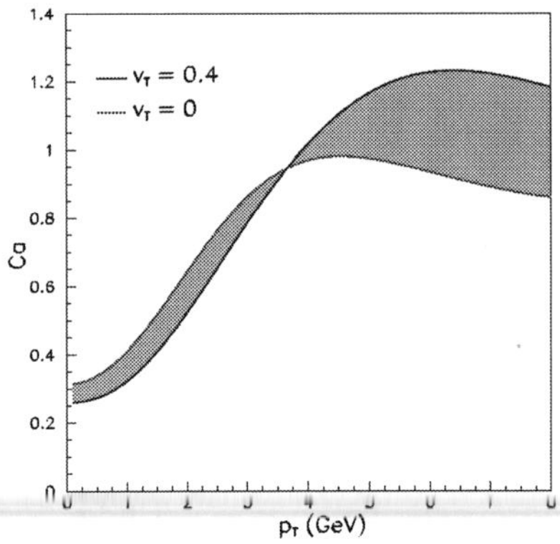

FIGURE 1. Charmed baryon to meson ratio, Ca, as a function of transverse momentum. The parameters used in the calculation are $m^B = 2.29$ GeV, $m^M = 1.87$ GeV, $\tau_0 = 1$ fm, $T_0 = 200$ MeV, $T_f = 100$ MeV, corresponding to a final time $\tau_f = 8$ fm. Shown is a range when varying the transverse expansion velocity v_T from 0 (upper curve at low p_T) to 0.4 (lower curve at low p_T).

flat up to $p_T \simeq 5$ GeV and consistent with the value reported in Ref. [19]. Therefore, for simplicity we take a as a constant equal to the above quoted number. Thus

$$T^e_{AA} \simeq \frac{(1+a^{\text{int}})}{(1+xa^{\text{int}})} \frac{1+x(dN^{\Lambda}_{AA}/dp_T)/(dN^{D}_{AA}/dp_T)}{1+(dN^{\Lambda}_{AA}/dp_T)/(dN^{D}_{AA}/dp_T)}. \tag{23}$$

Figure 2 shows T^e_{AA} as a function of p_T. We have used a range of values for the transverse expansion velocity between $v_T = 0$ and $v_T = 0.4$. We see that for the chosen evolution parameters, T^e_{AA} is indeed smaller than 1 and thus it contributes to the suppression of the single non-photonic electron nuclear modification factor R^e_{AA}.

SPIN ALIGNMENT OF VECTOR MESONS

The study of spin polarization of produced hadrons in reactions at high energies has opened a window to the understanding of the underlying dynamics of quark recombination. In the context of heavy-ion collisions, polarization studies can also help to understand the evolution of the system from its early stages [33, 34, 35].

Polarization analyses require to determine a given direction that serves as the spin quantization axis. From the experimental point of view, it is possible to determine two directions: the normal to the reaction and the normal to the production planes. The fist plane is defined as the one containing the impact parameter and the beam direction vectors whereas the second one is defined as containing the hadron's final momentum and the beam direction vectors.

Polarization studies have long been carried out for hyperons at different beam energies using different kinds of colliding systems. The available data on Λ and $\bar{\Lambda}$ polarization in A + A collisions at RHIC energies is reported in Ref. [36]. These polarization

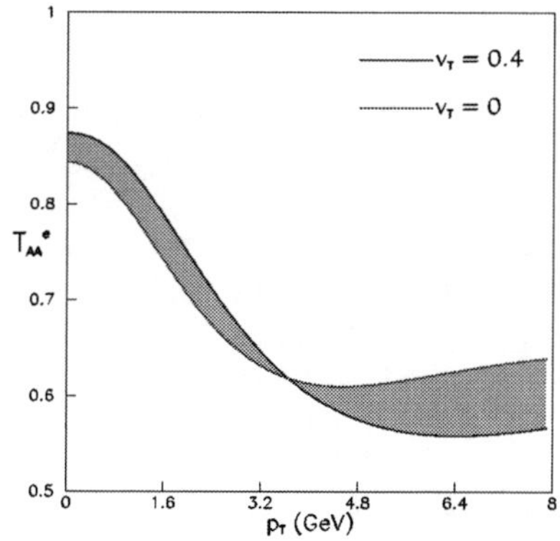

FIGURE 2. Suppression factor T^e_{AA} as a function of transverse momentum. The parameters used in the calculation are $m^B = 2.29$ GeV, $m^M = 1.87$ GeV, $\tau_0 = 1$ fm, $T_0 = 200$ MeV, $T_f = 100$ MeV, corresponding to a final time $\tau_f = 8$ fm, $x = 0.14$, a=0.073. Shown is a range for the transverse expansion velocity form $v_T = 0$ (lower curve at low p_T) and $v_T = 0.4$ (upper curve at low p_T).

measurements are referred to the reaction plane and are consistent with zero polarization. Other hyperon polarization measurements in A + A with respect to the production plane are only available at lower energies and for smaller systems [37] and report mainly only average values for the polarization but not its p_T dependence [38, 39]. In summary, the existing data on the hyperon p_T dependent polarization in A + A reactions is scarce.

On the other hand, the STAR collaboration has recently reported measurements of the p_T dependence of the polarization of the vector mesons ϕ and K^* [40]. These measurements refer to the 00 component of the so called *spin alignment* density matrix ρ, which is the density matrix for a two-spin one-half system in a triplet state, expressed in terms of the coupled basis [41]. Recall that a value $\rho^{00} = 1/3$, means that the spin of the vector meson is not aligned with respect to the chosen quantization axis. Deviations from this value indicate a degree of polarization of the vector spin which ultimately might reflect a polarization of the constituent quarks.

The experimental findings reported can be summarized as follows: When the spin alignment is referred to the *reaction plane* in Au + Au collisions at $\sqrt{s_{NN}} = 200$ GeV and measured at mid-rapidity, $\rho_{00} \simeq 1/3$, and it remains constant both as a function of p_T in the range $0 < p_T < 5$ GeV, for mid-central collisions, and as a function of the average number of participants in the same p_T range, for both ϕ and K^*. When the spin alignment is referred to the *production plane* and measured at mid-rapidity, both for p + p and mid-central Au + Au collisions at $\sqrt{s_{NN}} = 200$ GeV, $\rho_{00} > 1/3$ and it has a concave shape as a function of p_T in the range $0 < p_T < 5$ GeV with minima at slightly different intermediate values of p_T for ϕ and K^*.

An interesting observation that can be inferred from the above listed results is that the dynamics of hadron formation seems to play a role to orient the spin of quarks that form meson.

A possible origin of a quark polarization driving the polarization of vector mesons is discussed in Refs. [33] which study the transfer of local relative angular momentum in peripheral nuclear collisions to the quark spin polarization by means of rescattering during the reaction. However, this mechanism also predicts a small global polarization growing almost linearly with impact parameter which, according to the aforementioned results is not observed in data.

Another interesting possibility that has only been studied in the context of hyperon polarization [42, 43] is the scenario where the spin of a quark is oriented during the recombination process. The semiclassical picture accounting for the polarization is the Thomas precession produced by the accelerating force that pulls a slow moving quark (q^s) to form a fast moving hadron [42]. This mechanism also predicts that if the quark is fast (q^f) and is decelerated to form the hadron, its polarization will be of opposite sign compared to the case when it is accelerated.

Recall that recombination is a main channel for hadron production spanning the intermediate p_T region $2 \lesssim p_T \lesssim 5$. In this range, one can assume that the formed hadron is made up from the recombination of a slow and a fast quark. The polarization of the resulting hadron can be thus used as a testing ground of such scenario. Other scenario where fragmentation of a fast quark, instead of recombination is proposed as the main mechanism to produce the vector mesons is discussed Ref. [44], For this model to work, the authors assume, based on the e+ e- annihilation data, that the polarization of the anti-quark in the meson is proportional to that of the fragmenting one but opposite in sign. In contrast, as stated above, this is a natural consequence of Thomas precession in the recombination of slow and fast quarks.

For the Thomas precession to work, the pulling force is required to not be parallel to the original quark velocity since the Thomas frequency is a vector formed by the cross product of the force **F** and this velocity β, namely

$$\omega_T = \left(\frac{\gamma}{1+\gamma}\right) \mathbf{F} \times \beta, \tag{24}$$

where γ is the Lorentz gamma-factor. The polarization is given by [42]

$$\mathscr{P}^{s,f} = \mp \frac{\omega_T^{s,f}}{\Delta E}, \tag{25}$$

where the $-$ $(+)$ sign refers to the q^s (q^f). $\omega_T^{s,f}$ is the magnitude of the Thomas precession frequency for q^s and q^f, respectively and ΔE is the change of energy in the process of hadron formation.

Here I give a brief account on how the Thomas spin precession mechanism can describe the spin alignment of vector mesons produced at central rapidity in Au + Au and p + p collisions at $\sqrt{s_{NN}} = 200$ and show that under very simple assumptions, data for ρ_{00} are well reproduced within this approach. Details of the calculation can be found in Ref. [45].

The physical picture is that of a fast quark that decelerates and a slow one that accelerates to form a fast moving hadron. In the process, Thomas precession makes the spin of the former to acquire a positive polarization whereas the latter acquires a negative one. For central rapidities, the large momentum component of the hadron will thus be its transverse momentum p_T^H whereas the small component will be its longitudinal one, $p_{\parallel}^H$. We will assume that in the beam collision, a hard interaction produces a fast quark moving with a large transverse momentum p_T^f and, to simplify matters, a vanishing longitudinal momentum. This fast quark combines with the slow one, that we assume moves originally mainly in the longitudinal direction with momentum $p_{\parallel}^s$ and, also for simplicity, take it with vanishing transverse momentum. This sharp difference in the original direction of motion of the recombining quarks is at the core of the produced polarization since, as we proceed to show, it gives rise to a distinct p_T dependence of ρ_{00} which seems to be also observed in data.

In order to form the hadron, which should move with an intermediate value of momentum, between that of the q^f and of the q^s, the fast quark should slow down whereas the slow quark should speed up. Notice that for this mechanism to work, there is no need to assume that the process happens only in either a proton-proton or a nucleus-nucleus collision. Notice also that the momentum of the formed hadron provides a fixed direction to define that q^s (q^f) decelerates (accelerates), whereas, when referred to the reaction plane, no such fixed direction exists, since the direction of the impact parameter vector changes from one reaction to another and in such situation either quark can accelerate or decelerate.

The pulling force is equal to the change in momentum $\Delta\mathbf{p}$ of the given quark, in the interval of time Δt for the recombination process to happen, that is

$$\mathbf{F} = \frac{\Delta\mathbf{p}}{\Delta t}. \tag{26}$$

Thus ω_T for the given quark can be computed as the average over this time interval [42], namely

$$\begin{aligned} \omega_T^{s,f} &\propto \frac{\Delta p^{s,f}}{\Delta t}\beta^{s,f}\left(\int_{\Delta t} dt \sin\theta^{s,f}/\Delta t\right) \\ &\simeq \frac{\Delta p^{s,f}}{\Delta t}\beta^{s,f}\langle\sin\theta\rangle^{s,f}, \end{aligned} \tag{27}$$

where in the last line we have changed the time average of the sine of the angle by the average sine of the angle between the quark velocity vectors and their corresponding change in momentum. From Eqs. (25) and (27), we see that the calculation of the q^f and q^s polarizations reduces to computing the magnitudes of their change in momentum $\Delta p^{s,f}$, their $\langle\sin\theta\rangle^{s,f}$ and the change in energy ΔE.

In Ref. [45], it is shown that for the cross product of $\Delta\mathbf{p}^f$ with β^f one gets

$$\begin{aligned} \Delta\mathbf{p}^f \times \beta^f &= p_{\parallel}^{f/H}\beta^f\langle\sin\theta\rangle^f \\ &= (1-x_{\parallel})p_{\parallel}^H\langle\sin\theta\rangle^f, \end{aligned} \tag{28}$$

where we have introduced the definitions for the momentum fractions that the q^s has inside the hadron,

$$\begin{aligned} x_{\|} &= p_{\|}^{s/H}/p_{\|}^{H} \\ x_T &= p_T^{s/H}/p_T^{H}, \end{aligned} \tag{29}$$

and neglected the longitudinal hadron's momentum with respect to its transverse one, enforced momentum conservation $x_T + x_{\|} = 1$ and approximated $\beta^f \simeq 1$.

To compute $\langle \sin\theta \rangle^s$, notice that given that initially β^s points along the parallel direction and $\Delta \mathbf{p}^s$ is basically directed along the perpendicular direction, the initial angle between these vectors is $\pi/2$ and the average one should be close to $\pi/4$. Thus $\langle \sin\theta \rangle^s \simeq 1/\sqrt{2}$. Rather than approximating β^s (and therefore γ^s), we introduce a factor a for this polarization and let this vary such that $0 < a < 1$. Thus we write

$$\begin{aligned} \omega^s &= a\frac{\Delta p^s}{\Delta t} \\ a &= \left(\frac{\gamma^s}{1+\gamma^s}\right)\beta^s \langle \sin\theta \rangle^s. \end{aligned} \tag{30}$$

To compute $\langle \sin\theta \rangle^f$, notice that since $\Delta \mathbf{p}^f$ and β^f are almost perpendicular, $\langle \sin\theta \rangle^f \simeq 1$.

The change in energy is common to both the accelerating q^s and the decelerating q^f

$$\begin{aligned} \Delta E &= \{[(p_T^f)^2 + (p_{\|}^f)^2 + (m^f)^2]^{1/2} + [(p_T^s)^2 + (p_{\|}^s)^2 + (m^s)^2]^{1/2} \\ &- [(p_T^H)^2 + (p_{\|}^H)^2 + (m^H)^2]^{1/2}\} \\ &\simeq \{[(p_T^f)^2 + (m^f)^2]^{1/2} + [(p_{\|}^s)^2 + (m^s)^2]^{1/2} - [(p_T^H)^2 + (m^H)^2]^{1/2}\} \\ &= \left\{ p_T^f \left[1 + \frac{(m^f)^2}{2(p_T^f)^2}\right] + [(p_{\|}^s)^2 + (m^s)^2]^{1/2} - p_T^H \left[1 + \frac{(m^H)^2}{2(p_T^H)^2}\right] \right\}, \end{aligned} \tag{31}$$

where we have set $p_T^s = p_{\|}^f = 0$, neglected $(p_{\|}^H)^2$ compared with $(p_T^H)^2$ and expanded the square roots assuming that the transverse momentum components are large. Introducing the relation between the hadron and q^f transverse momenta $p_T^f = p_T^H/z$, with $0 < z < 1$, we get

$$\begin{aligned} \Delta E &= \left\{ \frac{p_T^H}{z}\left(1 + \frac{z^2 (m^f)^2}{2(p_T^H)^2}\right) + \left[\left(\frac{m^s}{m_T^H}\right)^2 (p_{\|}^H)^2 + (m^s)^2\right]^{1/2} \right. \\ &- \left. p_T^H \left(1 + \frac{(m^H)^2}{2(p_T^H)^2}\right)\right\}, \end{aligned} \tag{32}$$

where we have made use of the assumption that the rapidity of the q^s coincides with that of the hadron. Notice that the above expression can be simplified. In particular, since

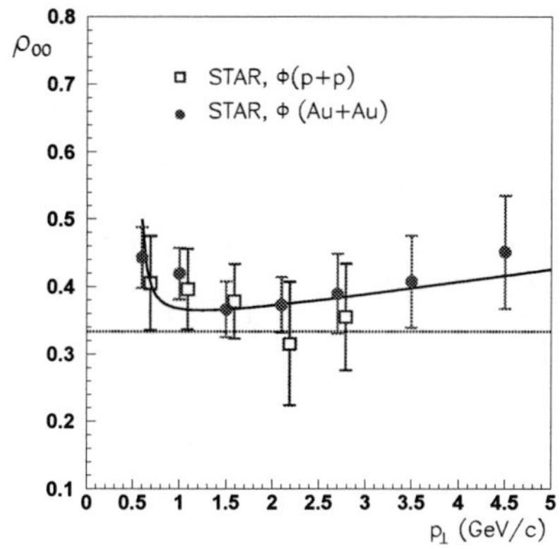

FIGURE 3. (Color online) ρ_{00} as a function of p_T for $\phi(1020)$ using the model parameters described in the text, compared to data from STAR [40] for p +p and Au + Au collisions at centrality 20-60 %, measured with respect to the production plane. For clarity, the p_T for p + p data has been displaced by 0.09 GeV with respect to the reported central value. The statistical and systematic errors have been added in quadrature. For comparison, we also draw the constant value 1/3 that represents the absence of polarization.

$p_\parallel^H = m_T^H \sinh y^H$, we can write

$$\left[\frac{(m^s)^2}{(m_T^H)^2}(p_\parallel^H)^2 + (m^s)^2\right]^{1/2} = m^s \cosh y^H. \tag{33}$$

Therefore, ΔE can be expressed as

$$\Delta E = \left\{\left(\frac{1-z}{z}\right) p_T^H + \left[\frac{z(m^f)^2 - (m^H)^2}{2p_T^H}\right] + m^s \cosh y^H\right\}. \tag{34}$$

We enphasize that the approximation to compute ΔE is such that Eq. (34) is valid for $p_T \gtrsim m^H$ and that for $z \lesssim 1$ the validity of the approximation can be extended to lower values of p_T.

It is now easy to compute the polarization for the slow and fast quarks by means of Eq. (25) and from them, the ρ_{00} density matrix element given by

$$\rho_{00} = \frac{1 - \mathscr{P}^s \mathscr{P}^f}{3 + \mathscr{P}^s \mathscr{P}^f}. \tag{35}$$

First we apply the model to compute ρ_{00} for ϕ. This case is the simplest one to treat within our approach since the quark content of this particle is $s\bar{s}$ and thus either one of these quarks can be thought of as being the fast (q^f) or the slow (q^s) one. Rather than making an exhaustive search in the parameter space, we choose reasonable values for them. We first fix the ϕ and strange quark masses to be $m^\phi = 1.02$ GeV, $m^{s,\bar{s}} = 0.5$ GeV.

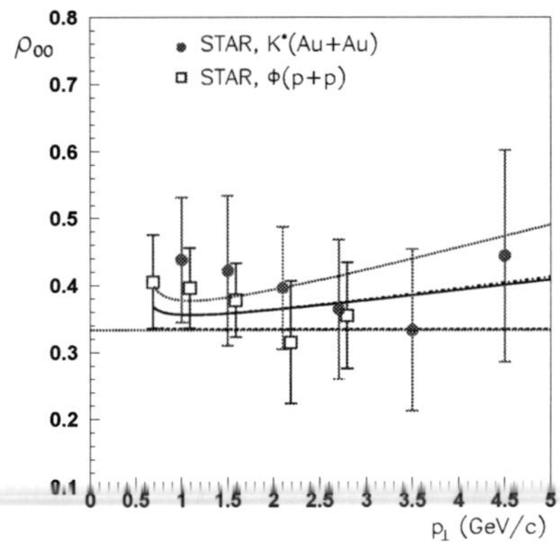

FIGURE 4. (Color online) ρ_{00} as a function of p_T for K^* using the model parameters described in the text, compared to data from STAR [7] for p +p and Au + Au collisions at centrality 20-60 %, measured with respect to the production plane. For clarity, the p_T for p + p data has been displaced by 0.09 GeV with respect to the reported central value. The statistical and systematic errors have been added in quadrature. The upper dashed curve represents the case for $\rho_{00}^{f=\mathrm{s},\, s=\mathrm{d}}$, whereas the lower dashed curve represents the case for $\rho_{00}^{f=\mathrm{d},\, s=\mathrm{s}}$. The intermediate solid curve represents $\rho_{00}^{K^*}$ as the algebraic average of the above. The intermediate dotted curve represents the case where ρ_{00} is computed using the average product of polarizations. For comparison, we also draw the constant value 1/3 that represents the absence of polarization.

The rapidity value we use is $y^H = 1$ and the formation time $\Delta t = 1$ fm. For the fractions of longitudinal and transverse momenta that the slow quark has inside the ϕ we take $x_T = x_\parallel = 0.5$. The fraction of the transverse momentum carried by the ϕ from the fast quark is taken as $z = 0.9$.

Figure 3 shows ρ_{00}^{ϕ} as a function of p_T^{ϕ} compared to data from STAR [7] for p + p and Au + Au collisions at centrality 20-60 %. A good description is obtained for $a = 0.25$.

Next, we apply this model to the case of K^*, whose quark content is d$\bar{\mathrm{s}}$. However, in this case one needs to be careful since the symmetry between the masses, present in the description of ϕ, is absent. Consequently the spin alignment has to be treated in average. To this end, let us first take a simple scenario and consider the arithmetic average in the following manner

$$\rho_{00}^{K^*} = \frac{1}{2}\left(\rho_{00}^{f=\mathrm{s},\, s=\mathrm{d}} + \rho_{00}^{f=\mathrm{d},\, s=\mathrm{s}}\right). \tag{36}$$

Figure 4 shows $\rho_{00}^{K^*}$ as a function of $p_T^{K^*}$ compared to data from STAR [7] for p + p and Au + Au collisions at centrality 20-60%, measured in the production plane. The curves are computed such that we employ the same set of parameters as in the computation of ρ_{00}^{ϕ} (with $m^{\mathrm{d}} = 0.3$ GeV) except for the value of z. The reason is that, whereas in the case that $q^f =$ s one can think that in order for this fast quark to pick up a d, the momenta of s and K^* are similar, when $q^f =$ d, its momentum must be larger, given the mass difference between K^* and d. Thus if $q^f = \mathrm{s}$ we choose $z = 0.9$ whereas when $q^f = \mathrm{d}$ we use $z = 0.3$. The upper dashed curve represents the case for $\rho_{00}^{f=\mathrm{s},\, s=\mathrm{d}}$, whereas the lower dashed curve is for $\rho_{00}^{f=\mathrm{d},\, s=\mathrm{s}}$. The intermediate solid curve represents $\rho_{00}^{K^*}$ as the algebraic average of the above, as in Eq. (36).

An alternative approach is to consider that the $\rho_{00}^{K^*}$ can be computed by the substitution

$$\mathscr{P}^s \mathscr{P}^f \to \frac{\mathscr{P}^{s=\mathrm{s}} \mathscr{P}^{f=\mathrm{d}} + \mathscr{P}^{s=\mathrm{d}} \mathscr{P}^{f=\mathrm{s}}}{2}, \tag{37}$$

that is, by the *average product of polarizations*, in Eq. (35). Figure 4 shows also this possibility represented by the intermediate dotted curve, using the same set of parameters for the cases were $q^f = \mathrm{s}$, $q^s = \mathrm{d}$ and $q^f = \mathrm{d}$, $q^s = \mathrm{s}$, as discussed above. As can be seen from the figure, no significant difference is found with the case where the average is taken with the functions ρ_{00} and both approaches give a good description of data.

CONCLUSIONS AND OUTLOOK

In this contribution I have studied two important phenomena that take place when final hadrons are mesured to have their transverse momentum in the intermediate ($1-2 \lesssim p_T \lesssim 4-5$ GeV) transverse momentum region, after a high energy collision between either protons or gold nuclei. These phenomena are: the anomalous suppression of the single, non-photonic electron yied and the spin alignment of vector mesons.

We have shown that the suppression of the nuclear modification factor R^e_{AA} can be partially understood by realizing that this quantity is affected by an enhancement in the charmed baryon to meson ratio at intermediate p_T in Au + Au collisions. We have argued that this enhancement is due to the fact that in this region, recombination of thermal light quarks with a heavy quark becomes a dominant mechanism for hadron production, in as much as recombination of thermal light quarks does to explain the proton to pion ratio enhancement. We have made use of the DQRM to calculate this ratio and have shown that for moderate and even for vanishing transverse expansion velocities, it indeed can be larger than the charmed baryon to meson ratio in p + p collisions. This enhancement in turn produces that the function T^e_{AA} is below 1 and thus contributes to the suppression factor introduced by considering energy losses due to the propagation of heavy flavors in the plasma.

Some important features concerning this result are worth keeping in mind: First, notice that we have not included the momentum shift introduced by energy losses when computing the transverse distributions of charmed mesons and baryons. This is so because for R^e_{AA}, energy losses should be included in the prefactor of the function T^e_{AA}. In this sense, in order to avoid a double counting of the effect, the ratio that goes into the calculation of this last function is the *raw* ratio. Second, it is expected that at some value of p_T, fragmentation becomes the dominant mechanism for hadron production and therefore that the charmed baryon to meson ratio decreases above that p_T value, given that fragmentation produces more mesons than baryons. Third, we have considered finite values of transverse flow for charmed mesons and baryons even thought it might be questionable that heavy flavors also flow as light flavors do. Nevertheless, there seems to be some experimental support for heavy quark flow [46]. In this sense, the flow strength range we have considered is only for moderate values. Notice however that even in the absence of flow the suppression factor keeps being less than 1.

We have also shown that data on the vector mesons ϕ and K^* spin alignment with respect to the production plane in Au + Au and p + p collisions are well described by assuming that these hadrons are produced by the recombination of a slow and a fast quark that in the process become polarized in opposite directions due to Thomas precession. In this plane, the momentum of the hadron provides a fixed direction to define whether a valence quark accelerates or decelerates. The mechanisms also sheds light on the fact that when no such fixed direction exists, the polarization vanishes. This is so because in this case, the impact parameter vector changes from one reaction to another and therefore so does the direction of the normal to the reaction plane. Thus the average polarization, which in the Thomas precession scenario involves the direction of the formed hadron, vanishes, since, when projecting this onto the normal to the reaction plane, it is proportional to the average of the cosine of the angle between these vectors, whose average vanishes. On the other hand, the polarization mechanism advocated in Refs. [33] relies on the existence of an initial correlation arising from the presence of an orbital angular momentum in peripheral collisions, which is then transferred to the quark spin through scattering. Since this angular momentum is related to the direction of the impact parameter vector, such relation provides the correlation to possibly make the above average to not vanish. Nevertheless, this mechanism does not explain why the constituent quarks forming a vector meson might have opposite polarizations casting some doubt on the dominance of this scenario for the intermediate p_T region.

We thus see that for not too fast quarks, either light or heavy, it is possible to understand certain features of final hadrons using the idea of a *pick up* reaction that allows the coalecense of these quarks with thermal ones. In the process, the final hadron populates the intermediate p_T region. The ideas discussed make use of a recombination scenario, which, in the case of the suppression factor for the single, non-photinic electrons nuclear modification factor, is explicit and, in the case of the spin alignment of vector mesons is only implicit. I believe that it is worth studing other phenomena in this momentum region such as the enhancement of purely strange baryons over purely strange mesons, the energy loss of not too fast quarks that form hadrons before fragmenting outside the quark-gluon plasma, etc. More detailed studies of these phenomena will help to clarify some of the surprissing results that have been dubed in the past as puzzels and contribute to better characterize the properties of the quark-gluon plasma.

ACKNOWLEDGMENTS

Support for this work has been received in part by DGAPA under PAPIIT grant IN116008.

REFERENCES

1. M. Gyulassy and X. N. Wang, Nucl. Phys. B **420**, 583 (1994); R. Baier, Y. L. Dokshitzer, A. H. Mueller, S. Peigne and D. Schiff, Nucl. Phys. B **484** (1997) 265; M. Gyulassy, P. Levai and I. Vitev, Nucl.

Phys. B **594**, 371 (2001); U. A. Wiedemann, Nucl. Phys. B **588**, 303 (2000); S. Wicks, W. Horowitz, M. Djordjevic and M. Gyulassy, Nucl. Phys. A **784**, 426 (2007).

2. For a recent review, see U. A. Wiedemann, arXiv:0908.2306 [hep-ph].
3. I. Arsene *et al.* [BRAHMS Collaboration], Nucl. Phys. A **757**, 1 (2005); B. B. Back *et al.*, Nucl. Phys. A **757**, 28 (2005); K. Adcox *et al.* [PHENIX Collaboration], Nucl. Phys. A **757**, 184 (2005); J. Adams *et al.* [STAR Collaboration], Nucl. Phys. A **757**, 102 (2005).
4. M. Gyulassy and L. McLerran, Nucl. Phys. A **750**, 30 (2005).
5. R. C. Hwa and C. B. Yang, Phys. Rev. C **67**, 034902 (2003); V. Greco, C. M. Ko and P. Lévai, Phys. Rev. Lett. **90**, 202302 (2003); R. J. Fries, B. Müller, C. Nonaka and S. A. Bass, *ibid.* **90**, 202303 (2003).
6. S. S. Adler *et al.* (PHENIX Collaboration), Phys. Rev. Lett. **96**, 032301 (2006).
7. B. I. Abelev *et al.* (STAR Collaboration), Phys. Rev. Lett. **98**, 192301 (2007); A. G. Knospe (STAR Collaboration), Eur. Phys. J. C **62**, 223 (2009).
8. M. Djordjevic, *et al.*, Phys. Lett. B **632**, 81 (2006).
9. N. Armesto, *et al.*, Phys. Lett. B **637**, 362 (2006).
10. S. Wicks, *et al.*, Nucl. Phys. **A784**, 426 (2007).
11. Y. L. Dokshitzer and D. E. Kharzeev, Phys. Lett. B **519**,199 (2001)
12. M. Djordjevic and M. Gyulassy, Nucl. Phys. **A733**, 265 (2004).
13. B.-W. Zhang, E. Wang and X.-N. Wang, Phys. Rev. Lett. **93**, 072301 (2004).
14. N. Armesto, C. A. Salgado and U. A. Wiedemann, Phys. Rev. D **69**, 114003 (2004).
15. M. G. Mustafa, Phys. Rev. C **72**, 014905 (2005)
16. M. Djordjevic, Phys. Rev. C **74**, 064907 (2006).
17. M. Djordjevic and U. Heinz. Phys. Rev. C **77**, 024905 (2008); Phys. Rev. Lett. **101**, 022302 (2008).
18. P. R. Sorensen and X. Dong, Phys. Rev. C **74**, 024902 (2006).
19. G. Martinez-Garcia, S. Gadrat and P. Crochet, *A new contribution to the nuclear modification factor of non-photonic electrons in Au+Au collisions at* $\sqrt{s_{NN}} = 200$ *GeV*, hep-ph/0702035; G. Martinez-Garcia, S. Gadrat and P. Crochet, Phys. Lett. B **663**, 55 (2008), *erratum* Phys. Lett. B **666**, 533 (2008).
20. S.S. Adler *et al.* (PHENIX Collaboration), Phys. Rev. C **69**, 034909 (2004).
21. J. Adams *et al.* (STAR Collaboration), *Measurements of identified particles at intermediate transverse momentum in the STAR experiment from Au+Au collisions at* $\sqrt{s_{NN}} = 200$ *GeV*, nucl-ex/0601042.
22. A. László and T. Schuster, (NA49 Collaboration), Nucl. Phys. A **774**, 473 (2006).
23. B.I. Abelev *et al.*, (STAR Collaboration), Phys. Lett. B **655**,104 (2007).
24. A. Ayala, M. Martinez, G. Paić and G. Toledo-Sánchez, Phys. Rev. C **77**, 044901 (2008).
25. Y. Oh and C. M. Ko, Phys. Rev. C **79**, 067902 (2009).
26. H. van Hees and R. Rapp, Phys. Rev. C **71**, 034907 (2005); H. van Hees, V. Greco and R. Rapp, Phys. Rev. C **73**, 034913 (2006); H. van Hees, M. Mannarelli, V. Greco and R. Rapp, Phys. Rev. Lett. **100**, 192301 (2008).
27. G. D. Moore and D. Teaney, Phys. Rev. C **71**, 064904 (2005).
28. A. Ayala, J. Magnin, L. M. Montaño and G. Toledo-Sánchez, Phys. Rev. C **80**, 064905 (2009).
29. C.J. Horowitz, E.J. Moniz and J.W. Negele, Phys. Rev. D **31**, 1689 (1985).
30. C. Horowitz and J. Piekarewicz, Nucl. Phys. A **536**, 669-696 (1992).
31. G. Toledo-Sánchez and J. Piekarewicz, Phys. Rev. C **65**, 045208 (2002).
32. F. Karsch, E. Laermann and a Peikert, Phys. Lett B **478**, 447 (2000); F. Karsch, Lect. Notes in Phys. **583**, 209 (2002).
33. A.-T. Liang and X.-N. Wang, Phys. Rev. Lett. **94**, 102301 (2005), Erratum, Phys. Rev. Lett. **96**, 039901(E) (2006); Phys, Lett. B **629**, 20, (2005); J.-H. Gao, S.-W. Chen, W.-t. Deng, Z.-T. Liang, Q. Wang and X.-N. Wang, Phys. Rev. C **77**, 044902 (2008).
34. B. Betz, M. Gyulassy, and G. Torrieri, Phys. Rev. C **76**, 044901 (2007).
35. F. Becattini, F. Piccinini, and J. Rizzo, Phys. Rev. C **77**, 024906 (2008).
36. B. I. Abelev, *et al.* (STAR collaboration), Phys. Rev. C **76**, 024915 (2007).
37. R. Bellweid, *et al.* (E896 Collaboration), Nucl. Phys. A698, 499c (2002).
38. J. W. Harris *et al.*, Phys. Rev. Lett. 47, 229 (1981)
39. A complete, though somewhat old review on the status of hyperon polarization in hadron systems can be found in A. D. Panagiotou, Int. J. Mod. Phys. A5, 1197 (1990).
40. B. I. Abelev *et al.* (STAR Collaboration), Phys. Rev. C **77**, 061902 (2008); J.H. Chen *et al.* (STAR Collaboration), J. Phys. G: Nucl. Phys. **35**, 044068 (2008); J.H. Chen *et al.* (STAR Collaboration), J.

Phys. G: Nucl. Phys. **34**, S331 (2007).
41. K. Schilling, P. Seyboth, and G. Wolf, Nucl. Phys. B **15**, 397 (1970); Erratum: Nucl. Phys. B **18**, 332 (1970).
42. T. A. DeGrand and H. I. Miettinen, Phys. Rev. D **24**, 2419 (1981).
43. A. Ayala, E. Cuautle, G. Herrera and L. M. Montaño, Phys. Rev. C **65**, 024902 (2002).
44. X. Qing-hua and L. Zuo-tang, Phys. Rev. D **68**, 034023 (2003).
45. A. Ayala, E. Cuautle, G. Herrera, J. Magnin and L. M. Montaño, Phys. Lett. B **682**, 408 (2010).
46. S. A. Butsyk (PHENIX Collaboration), Nucl. Phys. A **774**, 669 (2006).

Some scenarios with non-homogeneous Higgs vacuum

J. Lorenzo Diaz Cruz

Facultad de Ciencias Físico-Matemáticas, BUAP. Apdo. Postal 1364, C.P. 72000 Puebla, Pue., México

Abstract. The standard Higgs mechanism employed in the Standard Model (SM) for electroweak symmetry breaking, relies on a homogenous Higgs vacuum expectation value (v.e.v.), i.e. a vacuum that does not depend on the position or the time coordinates. However, other non-homogeneous structures could also be considered, either at long or short distances. For instance, spatial variations of the Higgs v.e.v. on cosmological scales, would induce variations of the fundamental constants, and are severely constrained. Other possibilities, such as a discrete microscopic structure of the Higgs vacuum, or a confined Higgs mechanism associated with a strongly interacting Higgs sector, could be tested and give some light on the electroweak-scale contributions to the cosmological constant.

Keywords: Electroweak, symmetry breaking, Higgs
PACS: 12.60.-i,11.10.Kk,95.35.+d

INTRODUCTION

Understanding the properties of the quantum vacuum is a central problem in contemporary physics, with implications that range from particle physics to cosmology. On the particle front, the notion of spontaneous symmetry breaking (SSB) [1] has been an important ingredient for the development of the standard model (SM), with applications that started with the description of chiral symmetry breaking in the strong interactions [2], and continued with the breaking of the electroweak gauge symmetry [3, 4]. Within the QCD dommain, the non-perturbative vacuum configuration is supposed to explain the appearence of light pions, while electroweak symmetry breaking (EWSB) is needed in order to generate the masses for the quarks, leptons and gauge bosons of the SM. Electroweak precision tests seem to prefer a light Higgs boson, with a mass of the order of the electroweak (EW) scale $v \simeq 254$ GeV . Plenty of phenomenological studies have provided us with an understanding of the expected properties of the Higgs boson (mass, decay rates and production cross sections), which should be tested soon at the LHC [7]. Applications of SSB in cosmology, include the development of inflationary model [5] and the generation of the baryon asymmetry [6].

However, the effective description of the Higgs mechanism still lacks a fundamental understanding. The need to understand the connection of the SM quantum vacuum with gravity, has been exacerbated by the observation of an accelerated expansion of the galaxies, which is usually interpreted in terms of a dark energy componnet for the matter-energy content of the universe. The above problems seem to suggest that there is a fundamental problem with the way we are interpreting SSB in quantum field theory.

CP1259, *First CINVESTAV-UNAM Symposium on High Energy Physics*
edited by J. C. D'Olivo, R. López-Fernández, and M. Á. Pérez

In this paper we want to discuss possible departures from the standard scenario of a continuous and homogeneous Higgs vacuum. Namely, we want to discuss alternative models for the structure of the vacuum. Such non-homogeneous structures could arise either at long or short distances. For instance, one could consider spatial variations of the Higgs v.e.v. on cosmological scales, which would induce variations of the fundamental constants. Other possibility is a discrete microscopic structure of the Higgs vacuum, which is motivated by the view that the Higgs vacuum could really be a Bose condensate. Such condensates have been studied in condensed matter, where some compounds are made of certain atoms, e.g. Helium, that favor the emergence of such phenomena. One could be tempted to extrapolate that such "molecular" or"atomic" structures should also exist in order to explain the true nature of the Higgs mechanism. Another possibility is the confined Higgs mechanism, which can be associated with a strongly interacting EWSB. These new approaches could also give some light on the cosmological constant problem.

In this paper, we discuss first (section 2) an scenario with spatial gradients for the vacuum expectation value (v.e.v.), which was analyzed in ref. [8]. Then in section 3 we shall consider the possibility that the vacuum has a discrete or foamy structure, which induces violations of Lorentz invariance [9]. Another scenario, where the Higgs v.e.v. is confined inside the quarks, leptons and gauge bosons, which was presented in ref. [10], is mentioned in section 4. A connection between the Higgs mechanism and the dark matter problem [13], is contained in section 5, while the conclusions appear in section 6.

COSMOLOGICAL VARIATIONS OF THE HIGGS VEV

Within the SM the 19 free parameters are really constant, i.e. they do not depend neither on the time nor on the location where the measurements are performed. In particular, the standard view of SSB assumes that the Higgs v.e.v. is homogeneous in space-time, which could seem the most natural scenario. However, there exist the logical possibility that different parts of the universe may have different values of the physical parameters. On one side we have the cosmological constant problem, which could require some form of anthropic selection. It could also be that our visible universe, where the values of the physical parameters should be constant, is only a small fraction of the total universe. One possible scenario considered in the literature [8], assume that there could be spatial variations of the Higgs vev, namely that it could have some spatial gradient. However, this scenarios is severely constrained, precisely by the fact that such variations would induce a huge change in the cosmological constant, as it was shown in ref. [8]. The largest change on the cosmological constant can be represented by the top quark contribution:

$$v\frac{\delta\Lambda}{\delta v} = m_t\frac{\delta\Lambda}{\delta m_t} = 10^{52}\Lambda_0 \tag{1}$$

It would be very hard to explain how the value of Λ_0 should be adjusted in order to get a value close to the experimental value.

A DISCRETE MODEL OF THE HIGGS VACUUM

Here we shall consider a model of the Higgs vacuum structure, which is asssumed to consist of small size regions (droplets) where the vacuum expectation value is different from outside; for simplicity we shall consider that these regions form spherical droplets. The model is not distinguishable from the SM at the energies of current accelerators, however interesting deviations can be expected to occur at the coming LHC or higher energies.

As we discussed in our paper [9], the continuum Higgs v.e.v. will be replaced by a distribution, i.e. we shall assume that the Higgs v.e.v. takes a certain value in some small regions (droplets), elsewhere the v.e.v. will have a different value. If the vacuum energy (v.e.v.) were spread continuosly, it would contribute to the cosmological constant, with a value of the order $\Lambda_{ew} \simeq 10^9$ GeV^4 = 10^{49} GeV/cm^3. However if the droplets, of finite size r_d and inter-distance l_d, are distributed uniformly with a density ρ_d, then their contribution to the cosmological constant would be $\Lambda_{ew} = \rho_d v \tau_d$, where the volume of the droplets is given by $\tau_d \simeq r_h^3$. Thus, by saturating the observed value for the cosmological constant ($\simeq 10^{-4} GeV/cm^3$), we obtain $\rho_d \tau_d \simeq 10^{-56}$. Furthermore, by considering that the distance between the droplets should be smaller than the shortest distance being tested at current colliders, i.e. $l_d \leq 10^{-15}$ cm, then the resulting size of the droplest is of the order of the Planck length, i.e. $r_d \simeq 10^{-33}$ cm.

One may wonder why current experiments have not detected the structure of the Higgs vacuum. The reason is that current probes (photons, electrons, protons) have an energy or momentum that corresponds to a wave-length that is large than the distance between the spheres with v.e.v. different from zero. Thus, with current probes the vacuum "looks" continuum. However, as one gets an energy that is if the order of the inverse of the distance between the spheres, the vacuum will start to show its structure.

In order to identify possible test of our model that can be carried at the LHC, we shall focus on Higgs phenomenology. Let us consider the standard Higgs interaction with a fermion ψ, which is described by the Yukawa lagrangian. After SSB we get the mass of the fermion and its interaction with the Higgs. However, this will be valid only at low energies, but at high-energies the fermion will "see" less v.e.v., therefore the coupling will be not be given just by the fermion mass, but rather we need to include an energy-dependent factor for the vertex and the mass:

$$\mathscr{L}_{new} = x(q^2)\frac{m}{v} h \bar{\psi}_L \psi_R + m(q^2) \bar{\psi}_L \psi_R + h.c. \tag{2}$$

These effect can be probed at the LHC by looking at the Higgs production. For instance we can study the gluon fusion production, which depends on the Higgs coupling with the top. Now the cross-section needs to include the form factor $x(q^2)$, which will affect the shape of the p_T distributions. At lower momenta the result will be similar to the SM, but at higher momenta, we will observe a deviation from the SM result.

A CONFINED HIGGS MECHANISM

In another proposal to understand the contribution from chiral symmetry breaking to the cosmolgical constant, Brodsky and Shrock [10] have studied the spatial dependence of the quark and gluon condensates, which are associated with chiral symmetry breaking. They argued that such condensates have support only within the hadrons. Therefore, one one should no follow the standard lore and adjudicate these condensates to the vacuum energy, but rather included them in the hadron masses. Thus, these condensates contribute in fact to the ordinary matter content of the universe. They also argue that if a similar mechanism were in fact generating the masses of quarks, leptons and gauge bosons, like some technicolor or composite Higgs scheeme, then the electroweak symmetry breaking would neither contribute to the cosmological constant, but rather to the usual matter content Ω_m.

DARK MATTER AND THE HIGGS MECHANSIM.

Plenty of astrophysical and cosmological data requires the existence of a DM component, that accounts for about 10-20% of the matter-energy content of our universe [11]. A weakly-interacting massive particle (WIMP), with a mass also of the order of the EW scale, seems a most viable option for the DM. What is the nature of DM and how does it fit into our current understanding of elementary particles, is however not known. Furthermore, given the similar requirements on masses and interactions for both particles, Higgs boson and DM WIMP, one can naturally ask whether they could share a common origin.

Within the minimal SUSY SM, which has become one of the most popular extensions of the SM, there are several WIMP candidates (neutralino, sneutrino, gravitino) [12]. Among them, the neutralino has been most widely studied; it is a combinations of SUSY partners of the Higgs and gauge bosons, the Higgsinos and gauginos. Thus, in SUSY models the fermion-boson symmetry provides a connection between the Higgs boson and DM. However, many new models have been proposed more recently, which provide alternative theoretical foundation to stabilize the Higgs mechanism. Another proposal for a dark matter candidate, was presented within the Holographic Higgs models. In these constructions, EWSB is triggered by a light composite Higgs boson, which emerges as a pseudo-goldstone boson [14, 15]. Within this class of models, we propose that a stable composite "Baryon", tightly bounded by the new strong interactions, can account for the DM [13]. Although, our DM candidate could share similar characteristics with other WIMPs, its composite nature will have important implications for cosmological bounds and the experimental searches for DM. This picture, where strong interactions produce a light pseudo-goldstone boson and a heavier stable fermion, is not strange at all in nature. This is precisely what happens in ordinary hadron physics, where the pion and the proton play such roles. We have considered models that produce a similar pattern for the Higgs and DM, but at a higher energy scale, and with a stable neutral state instead of a charged one. We believe that such scenario is very attractive and unifying, and it provides further understanding of both EWSB and DM problems.

CONCLUSIONS

Explaining the nature of electroweak symmetry breaking (EWSB) is one of the most important questions in particle physics today. Even if the Higgs bosons is found at the LHC, and one can determine what is the physics that explain the stability of the Higgs mechanism, and even if we identify the Dark matter candidate, there will be some issues left open. One of them, probably the most difficult one, is the cosmological constant problem. Namely, we would like to understand why the Higgs vacuum does not produce the large curvature that one would expect with naive estimates. Many efforts have been devoted to this problem, but so far no solution has been found.

Here, we have discussed other models of the Higgs vacuum, which departs from the usual continuum model. We have presented the consequences of a model of the Higgs vacuum with a spatial gradient holding at cosmological scales. Then, we have discussed a model where the vacuum contans small size regions where the vacuum expectation value is different from other regions. Finally, we have also discussed an scenario where the Higgs v.e.v. is confined to live only within the SM particles. These models are not distinguishable from the SM at the energies of current accelerators, however interesting deviations can be expected to occur at the coming LHC or at higher energies.

Recently, a general treatment of a foamy vacuum entering to the SM through the so called "Higgs portal", was presented in ref. [16], we are studying at present possible implications for our model [17].

ACKNOWLEDGMENTS

This paper is dedicated to the memory of our teacher, friend and collegue Augusto Gracia (QEPD). The author thanks the *Sistema Nacional de Investigadores* and *CONACyT* (México) for financial support.

REFERENCES

1. P. W. Higgs, Phys. Lett. **12**, 132 (1964).
2. For a review of SSB and chiral symmetry see the nobel lectures of Y. Nambu.
3. S. L. Glashow, Nucl. Phys. **22**, 579 (1961).
4. S. Weinberg, "A Model Of Leptons," Phys. Rev. Lett. **19**, 1264 (1967), A. Salam, "Weak And Electromagnetic Interactions," *Originally printed in *Svartholm: Elementary Particle Theory, Proceedings Of The Nobel Symposium Held 1968 At Lerum, Sweden*, Stockholm 1968, 367-377*
5. A. H. Guth, Phys. Rev. D **23**, 347 (1981).
6. B. Grinstein and M. Trott, Phys. Rev. D **78**, 075022 (2008) [arXiv:0806.1971 [hep-ph]].
7. For review of Higgs phenomenology see: J. Ellis, "Beyond the Standard Model at the LHC and Beyond," arXiv:0710.0777 [hep-ph]; J. L. Diaz-Cruz and O. A. Sampayo, Phys. Rev. D **50**, 6820 (1994).
8. J. F. Donoghue, JHEP **0303**, 052 (2003) [arXiv:hep-ph/0101130].
9. P. Amore, A. Aranda and J. L. Diaz-Cruz, arXiv:0808.0028 [hep-ph].
10. S. J. Brodsky and R. Shrock, arXiv:0905.1151 [hep-th].
11. G. Bertone, D. Hooper and J. Silk, Phys. Rept. **405**, 279 (2005) [arXiv:hep-ph/0404175].
12. J. R. Ellis, K. A. Olive, Y. Santoso and V. C. Spanos, Phys. Rev. D **71**, 095007 (2005) [arXiv:hep-ph/0502001]; J. L. Diaz-Cruz, J. R. Ellis, K. A. Olive and Y. Santoso, JHEP **0705**, 003 (2007) [arXiv:hep-ph/0701229].

13. J. L. Diaz-Cruz, Phys. Rev. Lett. **100**, 221802 (2008) [arXiv:0711.0488 [hep-ph]].
14. R. Contino, Y. Nomura and A. Pomarol, Nucl. Phys. B **671**, 148 (2003).
15. A. Aranda, J. L. Diaz-Cruz, J. Hernandez-Sanchez and R. Noriega-Papaqui, Phys. Lett. B **658**, 57 (2007) [arXiv:0708.3821 [hep-ph]].
16. G. F. Giudice, M. Raidal and A. Strumia, arXiv:1003.2364 [hep-ph].
17. We would like to thank communication with A. Strumia, regarding possible problems of our discrete Higgs model.

Radiative corrections in baryon semileptonic decays with the emission of a polarized baryon

C. Juárez-León*, A. Martínez*, M. Neri*, J.J. Torres† and R. Flores-Mendieta**

*Escuela Superior de Física y Matemáticas del IPN, Apartado Postal 75-702, México, D.F. 07738, Mexico
†Escuela Superior de Cómputo del IPN, Apartado Postal 75-702, México, D.F. 07738, Mexico
**Instituto de Física, Universidad Autónoma de San Luis Potosí, Álvaro Obregón 64, Zona Centro, San Luis Potosí, S.L.P. 78000, Mexico

Abstract. We present an overview of the calculation of radiative corrections to the Dalitz plot of baryon semileptonic decays with angular correlation between polarized emitted baryons and charged leptons. We discuss both charged and neutral decaying baryons, restricted to the three-body region of the Dalitz plot. Our analysis is specialized to cover two possible scenarios: The center-of-mass frames of the emitted and the decaying baryons. We have accounted for terms up to order $(\alpha/\pi)(q/M_1)^0$, where q is the momentum-transfer and M_1 is the mass of the decaying baryon, and neglected terms of order $(\alpha/\pi)(q/M_1)^n$ for $n \geq 1$. The expressions displayed are ready to obtain numerical results, suitable for model-independent experimental analyses.

Keywords: Radiative corrections, baryons
PACS: 14.20.Lq, 13.30.Ce, 13.40.Ks

INTRODUCTION

Professor Augusto García passed away on August 16, 2009. He was an outstanding physicist with interests in particle physics. Among other achievements, he made important contributions to the understanding of radiative corrections to integrated observables in hyperon semileptonic decays. This paper summarizes to some extent his latest contributions to the field.

The calculation of radiative corrections (RC) to the Dalitz plot of baryon semileptonic decays (BSD),

$$A \to B + \ell + \overline{\nu}_\ell, \tag{1}$$

is a rather formidable task. Previous work has been focused on computing these RC by considering non-vanishing polarizations of the decaying baryon A [1] and of the emitted baryon B [2, 3]. Here we outline the salient facts of these analyses.

There are six different charge assignments predicted by the light and heavy quark content of A and B in (1). Accordingly, the charged lepton ℓ can be an $e^{\pm}$, $\mu^{\pm}$, and even $\tau^{\pm}$. Thus, RC should be calculated so as to cover all these charge assignments. Besides, in order to keep experimental analyses model-independent it is necessary that RC be model-independent as well. Since RC depend on the form factors introduced in the uncorrected decay amplitude, it is also necessary that they can be organized into a form suitable for producing numerical results.

CP1259, *First CINVESTAV-UNAM Symposium on High Energy Physics*
edited by J. C. D'Olivo, R. López-Fernández, and M. Á. Pérez

The approach to get the model independence of RC is the generalization implemented for hyperons [4] of the treatment of virtual RC in neutron beta decay [5] and of the application of the Low theorem [6] to the bremsstrahlung RC developed in Ref. [7]. In order to cover all the charge assignments, one needs to know only the RC to the neutral decaying baryon (NDB) $A^0 \to B^+ + \ell^- + \bar{\nu}_\ell$ and to the charged decaying baryon (CDB) $A^- \to B^0 + \ell^- + \bar{\nu}_\ell$ cases. The RC for the other cases are obtained from these two [8].

This work is organized to cover two scenarios: The center-of-mass frames of the emitted and the decaying baryons. We start this overview by discussing the boundaries of the Dalitz plot in the center-of-mass frame of B. We will specialize our calculation to the three-body region of this plot. Then we turn to the model-independent calculation of virtual and bremsstrahlung RC. We then repeat the analysis for the case of the center-of-mass frame of the decaying baryon. We close with some concluding remarks.

DALITZ PLOT IN THE CENTER-OF-MASS FRAME OF THE EMITTED BARYON

To start with, let us consider the BSD

$$A^- \to B^0 + \ell^- + \overline{\nu}_\ell, \tag{2}$$

where the four-momenta and masses of A^-, B^0, ℓ^-, and $\overline{\nu}_\ell$ will be denoted by $p_1 = (E_1, \mathbf{p}_1)$, $p_2 = (E_2, \mathbf{p}_2)$, $l = (E, \mathbf{l})$, and $p_\nu = (E_\nu^0, \mathbf{p}_\nu)$, and by M_1, M_2, m, and $m_\nu = 0$, respectively. We first analyze the center-of-mass frame of B. Accordingly, $E_2 = M_2$ and $\mathbf{p}_2 = \mathbf{0}$ and all other variables are referenced to this frame hereafter. In passing, let us mention that the direction of a vector $\mathbf{p}$ will be denoted by a unit vector $\hat{\mathbf{p}}$ and whenever the expressions involved are not manifestly covariant, quantities like p_1, l, or p_ν will also denote the magnitudes of the corresponding three-momenta, unless stated otherwise.

The uncorrected transition amplitude M_0 for process (2) is given by [2]

$$\mathsf{M}_0 = \frac{G_V}{\sqrt{2}}[\overline{u}_B(p_2)W_\mu(p_1,p_2)u_A(p_1)][\overline{u}_\ell(l)O_\mu v_\nu(p_\nu)], \tag{3}$$

where u_A, u_B, u_ℓ, and v_ν are the Dirac spinors of the corresponding particles and

$$\begin{aligned} W_\mu(p_1,p_2) &= f_1(q^2)\gamma_\mu + f_2(q^2)\sigma_{\mu\nu}\frac{q_\nu}{M_1} + f_3(q^2)\frac{q_\mu}{M_1} \\ &\quad + \left[g_1(q^2)\gamma_\mu + g_2(q^2)\sigma_{\mu\nu}\frac{q_\nu}{M_1} + g_3(q^2)\frac{q_\mu}{M_1}\right]\gamma_5. \end{aligned} \tag{4}$$

Here $O_\mu = \gamma_\mu(1+\gamma_5)$, $q \equiv p_1 - p_2$ is the four-momentum transfer, and $f_i(q^2)$ and $g_i(q^2)$ are the weak vector and axial-vector form factors, respectively.

The observable effects of the polarization of B can be studied through the replacement

$$u_B(p_2) \to \Sigma(s_2)u_B(p_2), \tag{5}$$

where $\Sigma(s_2)$, the spin projection operator, reads

$$\Sigma(s_2) = \frac{1-\gamma_5 \not{s}_2}{2}, \tag{6}$$

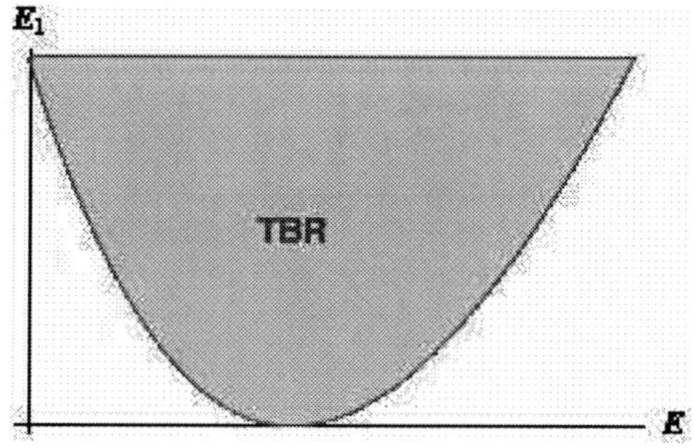

FIGURE 1. Kinematical region as a function of E and E_1 for BSD. The shadowed area is referred to as the three-body region (TBR).

with $s_2 \cdot s_2 = -1$ and $s_2 \cdot p_2 = 0$. In the center-of-mass frame of B, s_2 becomes the purely spatial unit vector $\hat{\mathbf{s}}_2$ which points along the spin direction.

The allowed kinematical region in the variables E and E_1 for process (2) is depicted in Fig. 1. The shadowed area is referred to as the three-body region (TBR) of the Dalitz plot and is bounded by [2].

$$m \leq E \leq E_m, \qquad \text{and} \qquad E_1^{\text{min}} \leq E_1 \leq E_1^{\text{max}}, \tag{7}$$

where

$$E_m = \frac{M_1^2 - M_2^2 - m^2}{2M_2}, \qquad \text{and} \qquad E_1^{\text{max,min}} = \frac{(M_2 + E \mp l)^2 + M_1^2}{2(M_2 + E \mp l)}. \tag{8}$$

Virtual radiative corrections

The virtual RC can be separated into a model-independent part M_v which is finite and calculable and into a model-dependent one which contains the effects of the strong interactions and the intermediate vector boson [5]. The latter can be accounted for in the definition of effective form factors f_1' and g_1' [4]. The decay amplitude M_V with virtual RC is thus given by

$$\mathsf{M}_V = \mathsf{M}_0' + \mathsf{M}_v, \tag{9}$$

where the prime on M_0 indicates that it contains the effective form factors explicitly, and

$$\mathsf{M}_v = \frac{\alpha}{2\pi}\left[\mathsf{M}_0\hat{\phi} + \mathsf{M}_{p_1}\hat{\phi}'\right], \tag{10}$$

with

$$\mathsf{M}_{p_1} = \left(\frac{E}{mM_1}\right)\frac{G_V}{\sqrt{2}}[\overline{u}_B W_\lambda u_A][\overline{u}_\ell \not{p}_1 O_\lambda v_\nu], \tag{11}$$

where the model-independent functions $\hat{\phi}(E)$ and $\hat{\phi}'(E)$ read [2]

$$\hat{\phi}(E) \;=\; 2\left[\frac{1}{\beta}\tanh^{-1}\beta - 1\right]\ln\left[\frac{\lambda}{m}\right] - \frac{1}{\beta}(\tanh^{-1}\beta)^2 + \frac{1}{\beta}L\left[\frac{2\beta}{1+\beta}\right]$$

$$+\frac{1}{\beta}\tanh^{-1}\beta-\frac{11}{8}+\begin{cases}\frac{\pi^2}{\beta}+\frac{3}{2}\ln\frac{M_2}{m} & \text{(NDB)}\\ \frac{3}{2}\ln\frac{M_1}{m} & \text{(CDB)}\end{cases} \tag{12}$$

and

$$\hat{\phi}'(E)=\left[\beta-\frac{1}{\beta}\right]\tanh^{-1}\beta, \tag{13}$$

Here $\beta\equiv l/E$, L is the Spence function and λ is the infrared-divergent cutoff.

After a standard procedure, the Dalitz plot with virtual RC can be expressed as [2]

$$d\Gamma_V=d\hat{\Omega}\left\{\hat{A}_0'+\frac{\alpha}{\pi}(\hat{A}_1'\hat{\phi}+\hat{A}_1''\hat{\phi}')-\hat{\mathbf{s}}_2\cdot\hat{\mathbf{l}}\left[\hat{A}_0''+\frac{\alpha}{\pi}(\hat{A}_2'\hat{\phi}+\hat{A}_2''\hat{\phi}')\right]\right\}, \tag{14}$$

where

$$d\hat{\Omega}=\frac{1}{2}\left[\frac{M_2}{M_1}\right]^3\frac{G_V^2}{2}\frac{dEdE_1d\Omega_\ell d\varphi_1}{(2\pi)^5}2M_2. \tag{15}$$

The functions $\hat{A}_0'$ and $\hat{A}_0''$ are given by

$$\hat{A}_0'=EE_\nu^0\hat{Q}_1-Ep_1(p_1-ly_0)\hat{Q}_2-l(l-p_1y_0)\hat{Q}_3-E_\nu^0p_1ly_0\hat{Q}_4+p_1^2ly_0(p_1-ly_0)\hat{Q}_5, \tag{16}$$

and

$$\hat{A}_0''=Ep_1y_0\hat{Q}_6+El\hat{Q}_7, \tag{17}$$

where $y_0=(p_1^2+l^2-E_\nu^{0^2})/(2p_1l)$ and by energy conservation $E_\nu^0=E_1-E-M_2$.

The $\hat{Q}_i$, which are quadratic functions of the form factors, can be found in Ref. [2]. On the other hand, the functions $\hat{A}_1'$, $\hat{A}_1''$, $\hat{A}_2'$, and $\hat{A}_2''$ that arise from virtual RC read

$$\hat{A}_1'=D_1EE_\nu^0-D_2l(l-p_1y_0), \qquad \text{and} \qquad \hat{A}_1''=\frac{M_2}{M_1}EE_\nu^0D_1, \tag{18}$$

$$\hat{A}_2'=E(l-p_1y_0)D_3-lE_\nu^0D_4, \qquad \text{and} \qquad \hat{A}_2''=\frac{EM_2}{M_1}(l-p_1y_0)D_3, \tag{19}$$

where the coefficients D_i are given by

$$D_1=f_1'^2+3g_1'^2, \qquad D_2=f_1'^2-g_1'^2, \tag{20}$$

$$D_3=2(-g_1'^2+f_1'g_1'), \qquad D_4=2(g_1'^2+f_1'g_1'). \tag{21}$$

Bremsstrahlung radiative corrections

Let us now discuss the four-body decay

$$A^-\to B^0+\ell^-+\overline{\nu}_\ell+\gamma, \tag{22}$$

where γ is a real photon with 4-momentum $k=(\omega,\mathbf{k})$ and which, in order to regulate the infrared divergence, is emitted with mass λ and with an additional longitudinal degree of freedom. By using the theorem of Low [6, 7] we can obtain the bremsstrahlung RC in a model-independent way. Thus, the bremsstrahlung amplitude M_B is [2]

$$\mathsf{M}_B=\mathsf{M}_{B1}+\mathsf{M}_{B2}+\mathsf{M}_{B3}, \tag{23}$$

with

$$\mathsf{M}_{B1}=e\mathsf{M}_0\left[\frac{2l\cdot\varepsilon}{2l\cdot k+\lambda^2}+\frac{2p_1\cdot\varepsilon}{\lambda^2-2p_1\cdot k}\right], \tag{24}$$

and

$$\mathsf{M}_{B2}=\frac{eG_V}{\sqrt{2}}\varepsilon_\mu[\overline{u}_B W_\lambda u_A][\overline{u}_\ell\frac{\gamma_\mu \not{k}}{2l\cdot k+\lambda^2}O_\lambda v_\nu]. \tag{25}$$

M_{B1} and M_{B2} contain terms of orders $1/k$ and $(k)^0$, respectively. As for M_{B3}, it can be ignored because its contribution to the decay rate is of order q/M_1 [1].

Again, after a standard calculation (see Ref. [2] for further details), the bremsstrahlung differential decay rate $d\Gamma_B$ corresponding to the TBR is

$$d\Gamma_B=d\Gamma'_B-d\Gamma_B^{(s)}, \tag{26}$$

where $d\Gamma'_B$ denotes half of the unpolarized decay rate whereas $d\Gamma_B^{(s)}$ contains the spin of the emitted baryon. We thus have[1]

$$d\Gamma'_B=d\Omega_1\left\{\frac{\beta^2(1-k^2x^2/\omega^2)}{(\omega-\beta kx)^2}(D_1EE_\nu+D_2\mathbf{l}\cdot\mathbf{p}_\nu)+\frac{1}{E(\omega-\beta kx)}\left[D_1E_\nu\left(\omega+2E-\frac{m^2\omega}{E(\omega-\beta kx)}-\frac{E(\omega-\beta kx)}{\omega}\right)+D_2\left[\mathbf{k}\cdot\mathbf{p}_\nu\left(1-\frac{m^2}{E(\omega-\beta kx)}+\frac{E}{\omega}\right)+\mathbf{l}\cdot\mathbf{p}_\nu\right]\right]\right\}, \tag{27}$$

and

$$\begin{aligned}d\Gamma_B^{(s)} &= -d\Omega_1(\hat{\mathbf{s}}_2\cdot\hat{\mathbf{l}})\left\{\frac{\beta^2(1-k^2x^2/\omega^2)}{(\omega-\beta kx)^2}(D_3E\hat{\mathbf{l}}\cdot\mathbf{p}_\nu+D_4E_\nu l)+\frac{1}{E(\omega-\beta kx)}\right.\\ &\quad\times\left[D_3\hat{\mathbf{l}}\cdot\mathbf{p}_\nu\left(\omega+2E-\frac{m^2\omega}{E(\omega-\beta kx)}-\frac{E(\omega-\beta kx)}{\omega}\right)\right.\\ &\quad\left.\left.+D_4E_\nu\left[\mathbf{k}\cdot\hat{\mathbf{l}}\left(1-\frac{m^2}{E(\omega-\beta kx)}+\frac{E}{\omega}\right)+l\right]\right]\right\},\end{aligned} \tag{28}$$

where

$$d\Omega_1=\frac{1}{2}\frac{\alpha}{\pi}\frac{G_V^2}{2}\frac{2M_2}{(2\pi)^6}\left[\frac{M_2}{M_1}\right]^3\frac{d^3p_1}{E_1}\frac{d^3l}{E}\frac{d^3k}{\omega}\frac{d^3p_\nu}{E_\nu}\delta^4(p_1-p_2-l-p_\nu-k). \tag{29}$$

[1] Let us notice that the effective form factors f_1' and g_1' may be used here, too. Again, this amounts to a rearrangement of order $(\alpha/\pi)^2$, valid within our approximations.

The expression for $d\Gamma_B^{(s)}$ has been rearranged in order to display the angular correlation $\hat{\mathbf{s}}_2 \cdot \hat{\mathbf{l}}$ explicitly. To achieve this we have used the replacement $\hat{\mathbf{s}}_2 \cdot \mathbf{p} \to (\hat{\mathbf{s}}_2 \cdot \hat{\mathbf{l}})(\hat{\mathbf{l}} \cdot \mathbf{p})$, with $\mathbf{p} = \mathbf{p}_1, \mathbf{k}, \mathbf{p}_\nu$, which is valid over the Dalitz plot after all other variables are integrated. Both expressions (27) and (28) contain the infrared divergence in their first summand within the curly brackets. The λ^2 that appears in Eq. (24) contributes in Eqs. (27) and (28) with linear and higher powers, which will become zero in the $\lambda^2 \to 0$ limit.

The integration over the neutrino variables with $\delta^3(\mathbf{p}_1 - \mathbf{l} - \mathbf{p}_\nu - \mathbf{k})$ in Eqs. (27) and (28) is trivial, but leaves a nontrivial argument inside the last $\delta(E_1 - M_2 - E - E_\nu - \omega)$, which allows to integrate over the photon momentum. The procedure is similar to the one implemented if Ref. [2] so details can be found there.

The partial results can be organized as

$$d\Gamma_B' = \frac{\alpha}{\pi} d\hat{\Omega} \left[\hat{A}_1' \hat{I}_0 + (\rho_1 + \rho_1')D_1 + (\rho_2 + \rho_2')D_2 \right], \tag{30}$$

and

$$d\Gamma_B^{(s)} = \frac{\alpha}{\pi} d\hat{\Omega} \hat{\mathbf{s}}_2 \cdot \hat{\mathbf{l}} \left[\hat{A}_2' \hat{I}_0 + (\rho_3 + \rho_3')D_3 + (\rho_4 + \rho_4')D_4 \right], \tag{31}$$

The integrals over the angular variables of the photon, $x = \hat{\mathbf{k}} \cdot \hat{\mathbf{l}}$ and φ_k, and over $y = \hat{\mathbf{p}}_1 \cdot \hat{\mathbf{l}}$ are left to be performed numerically, except in $\hat{I}_0$ which contains the infrared divergence and is obtained analytically. The functions $\hat{I}_0$, ρ_i and ρ_i' can be found in Ref. [2].

We have obtained the bremsstrahlung RC to the differential decay rate to order $(\alpha/\pi)(q/M_1)^0$. Let us now present the total differential decay rate by gathering together both virtual and bremsstrahlung contributions.

Final form of the radiative corrections

The complete Dalitz plot with the $\hat{\mathbf{s}}_2 \cdot \hat{\mathbf{l}}$ correlation to order $(\alpha/\pi)(q/M_1)^0$ restricted to the TBR and infrared convergent, $d\Gamma = d\Gamma_V + d\Gamma_B$, can be organized as

$$d\Gamma = d\hat{\Omega} \left\{ \hat{A}_0' + \frac{\alpha}{\pi} \hat{\Theta}_I - \hat{\mathbf{s}}_2 \cdot \hat{\mathbf{l}} \left[\hat{A}_0'' + \frac{\alpha}{\pi} \hat{\Theta}_{II} \right] \right\}, \tag{32}$$

with

$$\hat{\Theta}_I = \hat{A}_1'(\hat{\phi} + \hat{I}_0) + \hat{A}_1'' \hat{\phi}' + (\rho_1 + \rho_1')D_1 + (\rho_2 + \rho_2')D_2, \tag{33}$$

and

$$\hat{\Theta}_{II} = \hat{A}_2'(\hat{\phi} + \hat{I}_0) + \hat{A}_2'' \hat{\phi}' + (\rho_3 + \rho_3')D_3 + (\rho_4 + \rho_4')D_4. \tag{34}$$

Notice that the only difference between the NDB and the CDB cases is found in the function $\hat{\phi}(E)$, Eq. (12). The bremsstrahlung RC does not make any difference between both cases because of the order of approximation in this work. Let us also notice that Eq. (32) has triple integrals over some angular variables ready to be performed numerically. It requires that the numerical integrals in the RC be calculated within a Monte Carlo simulation every time E, E_1, and the form factors are varied, a task that represents a non-negligible computer effort. We shall now discuss a second form of the RC that should be more practical to use.

To proceed further, let us observe that for fixed values of E and E_1, Eqs. (33) and (34) take the form

$$\hat{\Theta}_r = a_r f_1^2 + b_r f_1 g_1 + c_r g_1^2, \tag{35}$$

because they are quadratic in the form factors. The subindex r takes the values $r = I, II$. Therefore, an alternative form of RC we propose consists in calculating arrays of the a_r, b_r and c_r coefficients determined at fixed values of (E, E_1) and that these pairs of (E, E_1) cover a lattice of points on the Dalitz plot.

To calculate the coefficients a_r, b_r and c_r it is not necessary to rearrange our final results to take the form (35). One can calculate them following a systematic procedure. One chooses fixed (E, E_1) points. Then one fixes $f_1 = 1$ and $g_1 = 0$, and obtains a_r, one repeats this calculation for $g_1 = 1$, $f_1 = 0$, to obtain c_r. Next, one repeats the calculation with $f_1 = 1$, $g_1 = 1$, and from these results one subtracts a_r and c_r, this way one obtains the coefficient b_r. The arrays of these coefficients should be fed into the Monte Carlo simulation. Within this simulation the repetitive triple integrations are reduced into a form of matrix multiplication.

We should remark that none of the forms of our RC results is compromised to fixing from the outset values for the form factors when such RC are applied in a Monte Carlo simulation. Let us now proceed to discuss the other reference system proposed above.

DALITZ PLOT IN THE CENTER-OF-MASS FRAME OF THE DECAYING BARYON

We now proceed to review the calculation when specializing to the center-of-mass frame of the decaying baryon, which has already been discussed in full in Ref. [3]. One might think that it could be possible to translate directly the final result of Ref. [2] into the final result of the latter RC by means of a Lorentz transformation. This, however, is not entirely possible so the calculation must be retaken starting at earlier stages.

We follow the same procedure as in the previous case and use the same conventions and notation, except for the fact now $p_1 = (M_1, 0, 0, 0)$.

The differential decay rate with virtual RC in the rest frame of A including the $\hat{\mathbf{s}}_2 \cdot \hat{\mathbf{l}}$ correlation, leaving to E and E_2 as the relevant variables, is

$$d\Gamma_V = d\Gamma_V' - d\Gamma_V^{(s)}, \tag{36}$$

which is valid for the two charge assignments $A^{\overline{0}}$. The unpolarized part has already been given in Ref. [9]; it reads

$$d\Gamma_V' = d\Omega \left[A_0' + \frac{\alpha}{\pi} \left(A_1' \phi + \phi' A_1'' \right) \right]. \tag{37}$$

The expressions for A_0', A_1', A_1'', ϕ, and ϕ' can be found in Eqs. (B1), (B2), (B3), (11), and (12) of this reference. The polarization dependent part is [3]

$$d\Gamma_V^{(s)} = d\Omega \left[A_0^{(s)} + \frac{\alpha}{\pi} \left(B' \phi + \phi' B'' \right) \right] \hat{\mathbf{s}}_2 \cdot \hat{\mathbf{l}}. \tag{38}$$

The trace calculation will lead to products $s_2 \cdot a$ with $a = l, p_\nu, p_1$. These products are specialized to the rest frame of A using the relation [3]

$$s_2 \cdot a = \mathbf{s}_2^A \cdot \left[\frac{\mathbf{p}_2}{E_2} a_0 - \mathbf{a} \right] = \hat{\mathbf{s}}_2^B \cdot \left[\frac{\mathbf{p}_2}{M_2} \left(a_0 - \frac{\mathbf{p}_2 \cdot \mathbf{a}}{E_2 + M_2} \right) - \mathbf{a} \right], \tag{39}$$

which corresponds to the Lorentz transformation from the rest frame of B to the rest frame of A. In the first equality of (39) it is understood that $\mathbf{p}_2$, $\mathbf{s}_2^A$, and the components of the 4-vector $a = (a_0, \mathbf{a})$ are specialized in the rest frame of A. In the second equality of (39) only $\hat{\mathbf{s}}_2^B$ still remains in the rest frame of B. The upper indices A and B on the spin of the emitted baryon emphasize this fact. In addition, the contribution of the correlations $\hat{\mathbf{s}}_2^B \cdot \hat{\mathbf{p}}_2$ and $\hat{\mathbf{s}}_2^B \cdot \hat{\mathbf{p}}_\nu$ to the $\hat{\mathbf{s}}_2^B \cdot \hat{\mathbf{l}}$ correlation over the Dalitz plot are taken into account with the substitution rule [9] $\hat{\mathbf{s}}_2^B \cdot \hat{\mathbf{p}} \to (\hat{\mathbf{s}}_2^B \cdot \hat{\mathbf{l}})(\hat{\mathbf{p}} \cdot \hat{\mathbf{l}})$, with $\hat{\mathbf{p}} = \hat{\mathbf{p}}_2, \hat{\mathbf{p}}_\nu$, which is valid under integration of the variables, other than E and E_2, contained in $d\Omega$. These steps will be extended to the bremsstrahlung part, where $\hat{\mathbf{p}} = \hat{\mathbf{k}}$ will also appear.

The results in Eq. (38) are given by [3]

$$A_0^{(s)} = Q_1' \left[p_2 y_0 \frac{M_1}{M_2} \right] + Q_2' \left[\frac{p_2 y_0}{M_2} \left(E - \frac{p_2 l y_0}{E_2 + M_2} \right) - l \right], \tag{40}$$

$$B' = -D_4 E_\nu^0 l + D_3 E (p_2 y_0 + l), \qquad \text{and} \qquad B'' = -D_3 E \hat{\mathbf{l}} \cdot \mathbf{p}_\nu, \tag{41}$$

where $E_\nu^0 = M_1 - E - E_2$ and $y_0 = (E_\nu^{0\,2} - p_2^2 - l^2)/2p_2 l$. The coefficients Q_1', Q_2', D_3 and D_4 can be found in the original work [3]. The primes on Q_1' and Q_2' indicate that the form factors f_1' and g_1' –where all the model-dependence has been absorbed– appear explicitly in these equations.

To the virtual RC of Eq. (37) one must add the bremsstrahlung RC. After a long but otherwise standard calculation, this part can be organized as

$$d\Gamma_B = d\Gamma_B' - d\Gamma_B^{(s)}, \tag{42}$$

where $d\Gamma_B'$ is independent of $\hat{\mathbf{s}}_2^B$ and can be identified with one-half the unpolarized decay rate of Eq. (27) of Ref. [9]. The B spin-dependent part is given by

$$d\Gamma_B^{(s)} = \frac{\alpha}{\pi} d\Omega \left(I_0 B' + D_3 \rho_3 + D_4 \rho_4 \right) \hat{\mathbf{s}}_2^B \cdot \hat{\mathbf{l}}. \tag{43}$$

$I_0(E, E_2)$ is the infrared-divergent integral and together with B' they are given in Ref. [9]. The contributions which are different with respect to the corresponding ones of Ref. [9] are

$$\begin{aligned} \rho_3 = {} & \frac{p_2 l}{4\pi} \int_{-1}^{1} dx \int_{-1}^{y_0} dy \int_0^{2\pi} \frac{d\phi_k}{D(1-\beta x)} \left\{ (p_2 y + l + \omega x) \left[\frac{\beta^2 (1-x^2)}{1-\beta x} + \frac{\omega}{E} \right] \right. \\ & \left. + \frac{\beta^2 (1-x^2)}{1-\beta x} \left[x E - \frac{D}{\beta} \right] \right\}, \end{aligned} \tag{44}$$

and

$$\rho_4 = \frac{p_2 l}{4\pi}\int_{-1}^{1}dx\int_{-1}^{y_0}dy\int_0^{2\pi}\frac{d\phi_k}{D(1-\beta x)}\left\{\left[\frac{\beta^2(1-x^2)}{1-\beta x}\right]l + E_\nu\left[-\beta+\left(\frac{1-\beta^2}{1-\beta x}-1-\frac{\omega}{E}\right)x\right]\right\}, \tag{45}$$

where

$$D = E_\nu^0 + (\mathbf{p}_2+\mathbf{l})\cdot\hat{\mathbf{k}}, \qquad \text{and} \qquad \omega = \frac{p_2 l(y_0-y)}{D}. \tag{46}$$

The neutrino energy in the presence of the real photon is $E_\nu = E_\nu^0 - \omega$, where $E_\nu^0 = M_1 - E_2 - E$. In Eqs. (44) and (45) $\mathbf{k}$ is referred to a coordinate axis system where $\hat{\mathbf{l}}$ points along the z-direction and $\hat{\mathbf{p}}_2$ lies on the (x,z) plane. The integration over $\mathbf{k}$ is performed with the variables $y = \hat{\mathbf{p}}_2\cdot\hat{\mathbf{l}}$, $x = \hat{\mathbf{k}}\cdot\hat{\mathbf{l}}$, and the azimuthal angle ϕ_k. They are ready to be performed numerically. However, their corresponding analytical parts are

$$\frac{2\rho_3}{p_2} = \frac{\zeta_{21}}{2E} + E^2(Y_2-Y_3) - 2\theta_0 E + Z_1 + \frac{1}{2}m^2[2(1-\beta^2)\theta_2 - 5\theta_3] + \frac{1}{2}(3E^2-2l^2)\theta_4 - \frac{3}{2}El\theta_5 - (1-\beta^2)\frac{E}{2}\theta_6 + \frac{3}{2}E\theta_7 + \frac{\theta_9}{4} + l^2\theta_{10} - \frac{l}{2}\theta_{14} - \frac{1}{2}(4E+E_\nu^0)\eta_0, \tag{47}$$

$$\rho_4 = \frac{p_2}{2}\left[l^2 Y_2 - \frac{1}{2}(2E-E_\nu^0)\eta_0 - \frac{1}{2}(E+2E_\nu^0)\gamma_0 + \frac{Y_4 E}{2} + \frac{l^2}{2}\theta_3\right]. \tag{48}$$

The explicit forms of the functions θ_i, γ_0, η_0, ζ_{ij}, Y_i and Z_1 are all found in Ref. [10].

Collecting the virtual and bremsstrahlung RC our final result is

$$d\Gamma\left(A^- \to B^0 e^- \overline{\nu}\right) = d\Omega\left\{\left[A_0' + \frac{\alpha}{\pi}\Theta_I\right] - \left[A_0^{(s)} + \frac{\alpha}{\pi}\Theta_{II}\right]\hat{\mathbf{s}}_2^B\cdot\hat{\mathbf{l}}\right\}, \tag{49}$$

where the explicit forms of Θ_I and A_0' coincide with Φ_1 of Eq. (54) and A_0' of Eq. (B1) of Ref. [9], respectively. Our new results are $A_0^{(s)}$ of Eq. (40) and

$$\Theta_{II} = B'[\phi + I_0(E,E_2)] + B''\phi' + D_3\rho_3 + D_4\rho_4, \tag{50}$$

where all the entries are defined above. The final result is infrared convergent.

The structure of RC has been organized in such a way that its practical application is straightforward, is a similar manner to the previous case. We will not discuss it further.

CONCLUDING REMARKS

The final results for the RC to the Dalitz plot of BSD with the angular correlation $\hat{\mathbf{s}}_2\cdot\hat{\mathbf{l}}$ between the polarization of the emitted baryon and the direction of the charged lepton are given, specialized to the center-of-mass frame of the emitted baryon by Eq. (32), and specialized to the center-of-mass frame of the decaying baryon by Eq. (49). These expressions are valid to order $(\alpha/\pi)(q/M_1)^0$ and cover the TBR of the Dalitz plot. They are model independent, they can be used in all charged assignments in different BSD (the charged lepton may be $e^{\pm}$, $\mu^{\pm}$, or $\tau^{\pm}$), and they are not compromised to predetermined values of the form factors of the uncorrected decay amplitude. And last but not least, they provide a good approximation to RC of medium- (several tens of thousands) and low- (several thousands) statistics experiments in light- and heavy-quark baryon semileptonic decays, respectively.

The resulting expressions may be used in a Monte Carlo simulation of an experimental analysis. However, performing the triple integrations every time E, E_1, and the form factors are varied may represent a very heavy computer effort. A more practical use of our result is through Eq. (35). Numerical arrays of the RC to the coefficients of the quadratic products of form factors may be first obtained at a lattice of points (E, E_1) covering the Dalitz plot and afterwards be fed in the Monte Carlo simulation. The computer effort within it would then be reduced to a sort of matrix multiplication.

The procedure outlined in this paper may be followed in the next future to extend the calculation of RC to BSD with the observation of polarization of the emitted baryon to cover the angular correlation $\hat{\mathbf{s}}_2 \cdot \hat{\mathbf{p}}_1$ and the four-body region. The precision of RC may be improved, while still preserving their model independence, by including terms of order $(\alpha/\pi)(q/M_1)$. It may be the case that experimental analyses should be limited to the center-of-mass frame of the decaying baryon A. Each one of these possibilities requires serious efforts which will be attempted elsewhere.

ACKNOWLEDGMENTS

The authors are grateful to Consejo Nacional de Ciencia y Tecnología (Mexico) for partial support. J. J. T. and A. M. were partially supported by Comisión de Operación y Fomento de Actividades Académicas (IPN). R. F.-M. was also partially supported by Fondo de Apoyo a la Investigación (UASLP).

REFERENCES

1. M. Neri, A. Martinez, A. Garcia, J. J. Torres and R. Flores-Mendieta, Phys. Rev. D **75**, 097301 (2007); Phys. Rev. D **74**, 077501 (2006), and references therein.
2. M. Neri, J. J. Torres, R. Flores-Mendieta, A. Martinez and A. Garcia, Phys. Rev. D **78**, 054018 (2008).
3. C. Juarez-Leon, A. Martinez, M. Neri, J. J. Torres, R. Flores-Mendieta and A. Garcia, Phys. Rev. D **79**, 057502 (2009).
4. A. Garcia and S. R. Juarez W., Phys. Rev. D **22**, 1132 (1980); **22**, 2923(E) (1980).
5. A. Sirlin, Phys. Rev. **164**, 1767 (1967).
6. F. E. Low, Phys. Rev. **110**, 974 (1958).
7. H. Chew, Phys. Rev. **123**, 377 (1961).
8. A. Martinez, J. J. Torres, A. Garcia and R. Flores-Mendieta, Phys. Rev. D **66**, 074014 (2002).
9. A. Martinez, J. J. Torres, R. Flores-Mendieta and A. Garcia, Phys. Rev. D **63**, 014025 (2000).
10. J. J. Torres, R. Flores-Mendieta, M. Neri, A. Martinez and A. Garcia, Phys. Rev. D **70**, 093012 (2004) [Erratum-ibid. D **75**, 019903 (2007)] and references therein.

Higgs masses and stability in the standard and the two Higgs doublet models

S.R. Juárez W.*, P. Kielanowski† and D. Morales C.*

*Departamento de Física, Escuela Superior de Física y Matemáticas, Instituto Politécnico Nacional, 07738 México D.F., Mexico
†Departamento de Física, Centro de Investigación y de Estudios Avanzados del IPN, 07360 México D.F., Mexico

Abstract. Within the framework of the standard model (SM) of elementary particles and the two Higgs doublet extension to this model (2DHM), we obtained analytical and numerical solutions for the gauge couplings, the vacuum expectation values (VEV) of the Higgs fields, the quark Yukawa couplings and quark masses, the quartic Higgs couplings, and the running Higgs masses, considering the renormalization group equations.The bounds on the SM Higgs running mass have been fixed, and the region of validity of the SM was determined through it, at the one and two loop approximations, using the triviality and stability conditions for the Higgs quartic coupling λ_H.

Keywords: Standard Model, 2 Higgs Doublet Model, Renormalization Group Equations
PACS: 11.10.Hi, 12.10.Kt, 12.60.Fr, 14.80.Bn

This paper is dedicated to the memory of our friend and collaborator Professor Augusto García whose work and passion has inspired so many around him.

THEORETICAL OVERVIEW

The questions about the mechanisms by which particles get their masses, and the relationship among different masses and forces of nature, are important and very basic to acquire an understanding of the nature of the constituents of matter and the forces among them.

Higgs found that parameters in the field equations associated with a scalar particle can be chosen in such a way that the lowest energy state of that field, the so called vacuum expectation value VEV, is different from zero. As the field is not zero in the empty space-time, all particles that interact with it gain mass from the interaction [1]. The Higgs boson gives the mechanism by which particles acquire mass.

To constrain these ideas more rigorously and to obtain very significant clues, it is very important to find evidence for the existence of Higgs field at first hand. The discovery of the Higgs particle is the most exciting prospect in contemporary particle physics [2]. To phrase *J.F. Gunion, et.al.* in [3]

> 'The Higgs sector of the Standard Model is not understood from a fundamental point of view, although it performs technically in an entirely satisfactory way as an effective low energy theory without conflict or contradiction. The physics that underlies electroweak symmetry breaking is simply

CP1259, *First CINVESTAV-UNAM Symposium on High Energy Physics*
edited by J. C. D'Olivo, R. López-Fernández, and M. Á. Pérez

not certain, although theorist may have guessed how it functions'.

THE HIGGS SECTOR OF THE STANDARD MODEL (SM)

The Standard Model (SM) is a quantum gauge field theory that describes the electroweak and strong interactions in terms of a symmetry principle, based on the gauge invariance under the $SU(3)_C \times SU(2)_L \times U(1)_Y$, transformations [4]. It is consistent with all the experimental results. The g_i, $(i = 1,2,3)$ associated with each of the gauge group are the gauge couplings. The gauge invariance of the model requires massless gauge bosons and fermions. Particularly in the electro weak sector $SU(2)_L \times U(1)_Y$ the Higgs mechanism circumvents this constraint by introducing a complex scalar field Φ through which the particles of the model acquire mass without breaking the gauge invariance and maintaining the renormalizabilty of the theory.

The scalar part of the Lagrangian

Let us now focus our attention on the three components of the Lagrangian where the Higgs field participates [5]

$$\mathscr{L}_{Higgs} = \mathscr{L}_{Kin} + \mathscr{L}_Y - V$$

The invariance of the theory under the local gauge transformations is guaranteed in the kinetic part of the Lagrangian by the covariant derivatives

$$\mathscr{L}_{kin} = D_\mu \Phi^\dagger D^\mu \Phi = \left| \left(\partial_\mu - i\frac{g'}{2} B_\mu Y - i\frac{g}{2} W_\mu^\alpha \tau_\alpha \right) \Phi \right|^2$$

Here is where the gauge coupling parameters $g_1 = \sqrt{\frac{5}{3}} g'$, $g_2 = g$ and g_3 are explicitly involved and where the $W^\pm$ and Z^0 arise as physical particles [6]. The Yukawa interaction, where the quark masses appear after diagonalization is given by

$$\mathscr{L}_Y = \overline{Q}_L \widetilde{\Phi} Y^u u_R + \overline{Q}_L \Phi Y^d d_R + \text{h.c.},$$

with

$$Y^{u(d)} = \text{Diag}\left(Y_{u(d)}, Y_{c(s)}, Y_{t(b)}\right) = \left(U_{u(d)}\right)_L y_{u(d)} \left(U_{u(d)}\right)_R^\dagger .$$

Here $y_{u(d)}$ are the matrices of the Yukawa couplings of quarks and the Higgs boson and $Y^{u(d)}$ is the diagonal matrix obtained from $y_{u(d)}$ by the biunitary transformation of the quark field with the matrices $\left(U_{u(d)}\right)_{L(R)}$. The Cabbibo Kobayashi Maskawa CKM Matrix [7] (which is observable) arises in terms of the diagonalization matrices

$$V = (U_u)_L (U_d)_L^\dagger .$$

The scalar potential of the Higgs field contains the quadratic (μ^2) and quartic (λ_H) couplings:

$$V = -\mu^2 \left(\Phi\Phi^\dagger\right) + \frac{1}{2}\lambda_H \left(\Phi\Phi^\dagger\right)^2$$

for which

$$\Phi = \begin{pmatrix} \phi_1 + i\phi_2 \\ \phi_3 + i\phi_4 \end{pmatrix}$$

and

$$\phi_3 \rightarrow \frac{1}{\sqrt{2}}(\nu + h).$$

The value of ν lies at the minimum of the potential $V(\Phi)$

$$\left.\frac{\partial V}{\partial \phi_3}\right|_{h=0} = -\mu^2\nu + \frac{1}{2}\lambda_H \nu^3 = 0$$

and the mass of the neutral Higgs boson is equal to

$$m_H^2 = \left.\frac{1}{2}\frac{\partial^2 V}{\partial h^2}\right|_{h=0} = \lambda_H \nu^2.$$

Thus the parameters of the potential are so constrained that the potential has a minimum and is bounded from below. This is known as the **Stability Condition** and it leads to

$$\lambda_H \geq 0. \tag{1}$$

If $h \neq 0$ and the field Φ has a non vanishing vacuum expectation value then we say that there is **Spontaneous Symmetry Breaking (SSB)**. The main consequence of the SSB is the generation of the masses for fermions and gauge bosons. The quark masses after the SSB are equal:

$$m_{u,c,t,d,s,b} = \frac{1}{\sqrt{2}}\nu Y_{u,c,t,d,s,b}.$$

RENORMALIZATION GROUP EQUATIONS

The Renormalization Group Equations (RGE) is a set of powerful tools [8], [9] that relate various observables (like couplings or masses) at different energies and also allow the study of their asymptotic behavior. In the perturbative quantum field theory the RGE are differential equations for the observables which are obtained from the condition that the S-matrix elements do not depend on the renormalization scheme or renormalization point. The right hand side of the RGE is an infinite series expanded according to the number of loops. Most of the numerical and analytical results for the RGE are to the order of one loop only possibly with a partial inclusion of two loops, while the RGE for most of the observables have been given for the SM and its extensions up to two loops [9].

The two loop renormalization group equations

The generic form of the renormalization group equation for the parameter x is the following

$$E\frac{dx}{dE}=\frac{dx}{dt}=\frac{1}{(4\pi)^2}\beta_x^{(1)}+\frac{1}{(4\pi)^4}\beta_x^{(2)},\quad t\equiv\ln\left(\frac{E}{\mu}\right).$$

In the Standard Model the renormalization group equations can be written for all the parameters present in the Lagrangian, i.e., g_i, $y_{u,d}$, λ_H and μ^2 or ν. The renormalization group flow of these parameters up to two loops is determined from the following equations

$$\begin{aligned}
\frac{dg_i}{dt} &= \frac{1}{(4\pi)^2}b_i g_i^3-\frac{g_i^3}{(4\pi)^4}\left(C_{if}\,\mathrm{Tr}(y_u^\dagger y_u)+\sum_k b_{ki}g_k^2\right) \qquad i=1,2,3,\\
\frac{dy_{u,d,e,\nu}}{dt} &= \frac{1}{(4\pi)^2}\left\{\beta^{(1)}_{u,d,e,\nu}+\frac{1}{(4\pi)^2}\beta^{(2)}_{u,d,e,\nu}\right\}y_{u,d,e,\nu},\\
\frac{d\lambda_H}{dt} &= \frac{1}{(4\pi)^2}\left\{\beta^{(1)}_{\lambda_H}+\frac{1}{(4\pi)^2}\beta^{(2)}_{\lambda_H}\right\},\\
\frac{d\ln v}{dt} &= \frac{1}{(4\pi)^2}\gamma_v^{(1)}+\frac{1}{(4\pi)^4}\gamma_v^{(2)},
\end{aligned}$$

where b_l are constants dependent on the model and G_l, $\beta^{(1)}_{u,d,e,\nu}$, $\beta^{(2)}_{u,d,e,\nu}$, $\gamma^{(1)}$, $\gamma^{(2)}$, $\beta_\lambda^{(1)}$, $\beta_\lambda^{(2)}$, are functions of the gauge couplings and the squares of the Yukawa couplings $H^{(1)}_{u,d,e,\nu}=y_{u,d,e,\nu}y^\dagger_{u,d,e,\nu}$, see Ref. [5] for the explicit values of the parameters for the SM and the 2DHM.

The one loop renormalization group equations

The one loop equations for the model parameters and the numerical values of the coefficients, are given in [10] and [11] for the SM and the 2DHM in the approximation in which the contribution of Y_d can neglected in relation to the one of Y_t.

$$\begin{aligned}
\frac{dg_i(t)}{dt} &= \frac{1}{(4\pi)^2}b_i g_i^3(t),\\
\frac{dv_{u,d}}{dt} &= \frac{1}{(4\pi)^2}[\alpha_1^{v_{u,d}}(t)+\alpha_3^{v_{u,d}}\,\mathrm{Tr}(y_u\, y_u^\dagger)]v_{u,d},\\
\frac{dy_{u,d}}{dt} &= \frac{1}{(4\pi)^2}[\alpha_1^{u,d}(t)+\alpha_2^{u,d}y_u y_u^\dagger+\alpha_3^{u,d}\,\mathrm{Tr}(y_u y_u^\dagger)]y_{u,d},
\end{aligned}$$

and

$$\frac{d\lambda_H}{dt}=\frac{12}{(4\pi)^2}\left[\lambda_H^2(t)+b_\lambda(t)\lambda_H(t)+c_\lambda(t)\right],$$

where

$$b_\lambda(t) \equiv \left[|Y_t(t)|^2 - \frac{3}{4}\left(\frac{1}{5}g_1^2(t) + g_2^2(t)\right)\right],$$

$$c_\lambda(t) \equiv \frac{3}{16}\left(\frac{3}{25}g_1^4(t) + \frac{2}{5}g_1^2(t)g_2^2(t) + g_2^4(t)\right) - |Y_t(t)|^4.$$

Analytical solutions of the one loop RGE

The one loop analytical solutions for the the quark sector of the former equations, are given in [10] and [11].

$$g_i(t) = g_i^0\left(1 - \frac{2b_i\left(g_i^0\right)^2(t-t_0)}{(4\pi)^2}\right)^{-\frac{1}{2}}, \quad g_i^0 \equiv g_i(t_0),$$

$$v_{u,d}(t) = v_{u,d}(t_0)\sqrt{r''_{v_{u,d}}(t)h_m^{\alpha_3^{v_{u,d}}}}(t), \quad v_{u,d}^0 = v_{u,d}(t_0),$$

$$Y_{u,c}(t) = Y_{u,c}(t_0)\sqrt{r(t)}h_m^{\alpha_3^u}(t), \; Y_t(t) = Y_t(t_0)\sqrt{r(t)}\,h_m^{\left(\alpha_2^u+\alpha_3^u\right)}(t),$$

$$Y_{d,s}(t) = Y_{d,s}(t_0)\sqrt{r'(t)}h_m^{\alpha_3^d}(t), \; Y_b(t) = Y_b(t_0)\sqrt{r'(t)}\,h_m^{\left(\alpha_2^d+\alpha_3^d\right)}(t),$$

where

$$h_m(t) = \exp\left(\frac{1}{(4\pi)^2}\int_{t_0}^t Y_t^2(\tau)d\tau = \left(\frac{1}{1 - \frac{2\left(\alpha_2^u+\alpha_3^u\right)}{(4\pi)^2}(Y_t^0)^2\int_{t_0}^t r_g(\tau)d\tau}\right)^{\frac{1}{2\left(\alpha_2^u+\alpha_3^u\right)}}\right.$$

and

$$r(t) = \exp\left[\frac{2}{(4\pi)^2}\int_{t_0}^t \alpha_1^u(\tau)d\tau\right] = \prod_{k=1}^{k=3}\left[\frac{g_k^2(t_0)}{g_k^2(t)}\right]^{\frac{c_k}{b_k}},$$

$$r'(t) = \exp\left[\frac{2}{(4\pi)^2}\int_{t_0}^t \alpha_1^d(\tau)d\tau\right] = \prod_{k=1}^{k=3}\left[\frac{g_k^2(t_0)}{g_k^2(t)}\right]^{\frac{c'_k}{b_k}}$$

$$r''_{v_{u,d}}(t) = \exp\left[\frac{2}{(4\pi)^2}\int_{t_0}^t \alpha_1^{v_{u,d}}(\tau)d\tau\right] = \prod_{k=1}^{k=2}\left[\frac{g_k^2(t_0)}{g_k^2(t)}\right]^{\frac{c''_{k(u,d)}}{b_k}}$$

are auxiliary functions.

TRIVIALITY AND STABILITY

The SM Quartic Higgs coupling

The knowledge of the energy dependence of $\lambda_H(t)$ is essential to determine the Higgs mass. The equation for the quartic coupling λ_H is a non linear Riccati type equation [12]

$$\frac{d\lambda_H}{dt} = \frac{12}{(4\pi)^2}\left\{\lambda_H^2(t) + \left[|Y_t(t)|^2 - \frac{3}{4}\left(\frac{1}{5}g_1^2(t) + g_2^2(t)\right)\right]\lambda_H(t) + \frac{3}{16}\left(\frac{3}{25}g_1^4(t) + \frac{2}{5}g_1^2(t)g_2^2(t) + g_2^4(t)\right) - |Y_t(t)|^4\right\}.$$

To solve it, we express $\lambda_H(t)$ using the auxiliary function $W(t)$

$$\lambda_H(t) = -\frac{4\pi^2}{12}\frac{W'(t)}{W(t)}.$$

In terms of this new function we get the following second order linear differential equation

$$W'' + A(t)W' + B(t)W = 0 \tag{2}$$

in which $A(t)$ y $B(t)$ are known regular functions.

The solution for $\lambda_H(t)$ is given by

$$\lambda_H(t) = -\frac{(4\pi)^2}{12}\frac{W_1'(t) - \frac{12}{(4\pi)^2}\lambda_H(t_0)W_2'(t)}{W_1(t) - \frac{12}{(4\pi)^2}\lambda_H(t_0)W_2(t)} \tag{3}$$

where $W_1(t)$ and $W_2(t)$ are two independent solutions of Eq. (2) that fulfill the following initial conditions

$$W_1(t_0) = 1, \quad \frac{d}{dt}W_1(t)|_{t_0} = 0, \quad W_2(t_0) = 0, \quad \frac{d}{dt}W_2(t)|_{t_0} = 1,$$

From the Stability and Triviality conditions we know that $\lambda_H(t)$ should be positive and finite. Thus, the range of the allowed values of λ_H remains between any positive value and zero or a singular value of λ_H as a function of the energy t.

The zero value of $\lambda_H(t)$ is determined from the zero of the numerator of the solution (3) i.e., the value of t, where $\lambda_H(t)$ vanishes becomes the allowed energy limit due to the Stability Condition

$$W_1'(t) - \frac{12}{(4\pi)^2}\lambda_H^n(t_0)W_2'(t) = 0 \quad \Rightarrow \quad \lambda_H^n(t_0) = ((4\pi)^2/12)W_1'(t)/W_2'(t).$$

The pole of $\lambda_H(t)$ is determined by the zero value of the denominator, that corresponds to limit due to the Triviality Condition.

$$W_1(t) - \frac{12}{(4\pi)^2}\lambda_H^d(t_0)W_2(t) = 0 \quad \Rightarrow \quad \lambda_H^d(t_0) = ((4\pi)^2/12)\,W_1(t)/W_2(t).$$

Considering the previous boundary values in the numerical evaluation of the quartic coupling and the Higgs mass at the two loop approximation we obtained [9] that the SM is valid in the whole range of energies for the following values of

$$0.368 \leq \lambda_H(m_t) \leq 0.621, \quad 150 \leq m_H \leq 193 \text{ GeV}.$$

SUMMARY OF THE RESULTS FOR THE SM PLATFORM

In [9] we studied the renormalization group evolution of the Higgs quartic coupling λ_H. The one loop equation for λ_H is non linear and it is of the Riccati type which we analytically and numerically solve in the energy range $[m_t, E_{GU}]$ where m_t is the mass of the top quark and $E_{GU} = 10^{14}$ GeV. We find that depending on the value of $\lambda_H(m_t)$ the solution for $\lambda_H(E)$ may have a singularity or a zero and become negative so the ultraviolet cut off of the standard model should be below or equal to the energy where the zero or singularity of λ_H occurs. We then numerically solve the two loop renormalization group equation for λ_H and compare it with the one loop solution. We find that the two loop running of λ_H is very sensitive to the evolution of the top quark Yukawa coupling Y_t. This implies a strong dependence on the top quark mass m_t and suggests that the choice of m_t as the renormalization point, that we use, reduces theoretical errors. We find that in the approximation of one loop for $0.397 \leq \lambda_H(m_t) \leq 0.618$ the Standard Model is valid in the whole range $[m_t, E_{GU}]$ while for two loops the bound is $0.368 \leq \lambda_H(m_t) \leq 0.621$. From the properties of λ_H we then study the predictions for the Higgs mass. We use the effective potential to derive the relation between the Higgs mass and λ_H and obtain that this relation is not very sensitive to the particular choice of the effective potential but for the large Higgs masses the two loop corrections are significant. We determine that the standard model is valid in the whole range $[m_t, E_{GU}]$ for the Higgs masses $153.5 \leq M_H \leq 191.1$ for one loop case and $148.5 \leq M_H \leq 193.1$ for two loops. The pattern of the behavior of $\lambda_H(E)$ for different values of $\lambda_H(m_t)$ indicates the existence of a phase transition in the standard model for $\lambda_H(m_t) = 0.5$ which corresponds to the value of the Higgs mass $M_H = m_t$.

THE 2DHM EXTENSION OF THE SM

The SM is so far supported by all the experimental results. Nevertheless the Higgs sector of the Standard Model (SM) requires further investigation in order to look for signals of new physics. In the SM the masses of the physical particles arise, after the spontaneous symmetry breaking (SSB), through its couplings with a single Higgs doublet. In the simplest extension of the SM, called the Two Higgs Doublet Model (2HDM), two Higgs doublets are introduced, (Φ_1, Φ_2), and a seven parameter potential for the Higgs fields. In this 2HDM model, the existence of five Higgs bosons is predicted. In the remaining section, we will focus our attention to the masses and the physical eigenstates of the new scalar particles: two charged ones ($H^{\pm}$) and three neutral (A^0),(h^0, H^0).

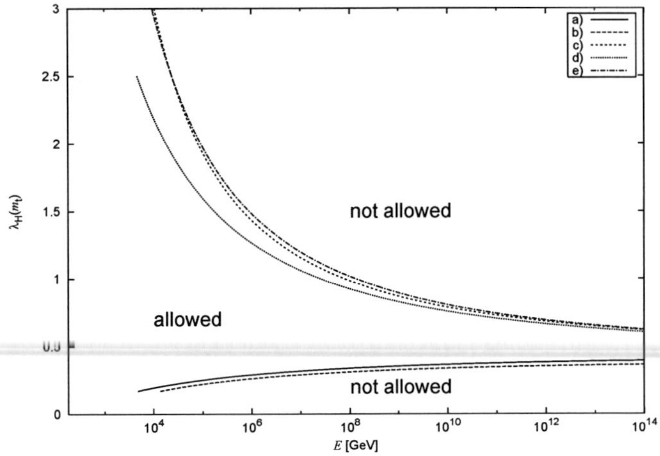

FIGURE 1. The range of the allowed values of the quartic coupling λ_H as the function of the energy for the Standard Model.

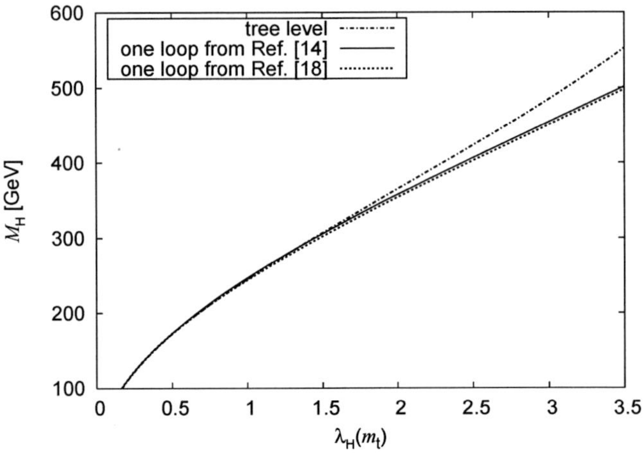

FIGURE 2. The relation between the $\lambda_H(m_t)$ and the physical Higgs mass the Standard Model. One can see that for large values of $\lambda_H(m_t)$ the relation becomes linear.

Masses in the two Higgs doublet extension of the SM

To simplify the analysis we have chosen, among different forms of the Lagrangian describing the same physical reality, a specific one in which the vacuum expectation values of both Higgs fields are real. We have obtained the masses of the physical particles and a set of conditions to be satisfied by the scalar parameters introduced in the potential, which determine the quartic couplings and self-couplings. In the 2HDM, we consider

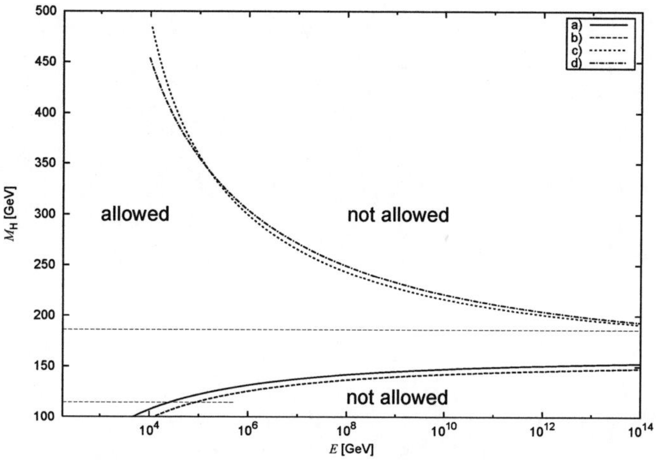

FIGURE 3. The range of the allowed values of the Higgs mass m_H as the function of the energy the Standard Model.

that the quark t obtains its mass through the VEV of first doublet $\langle\Phi_1\rangle_0 = v_1/\sqrt{2}$, and the quark b from the VEV of the second doublet $\langle\phi_2\rangle_0 = v_2/\sqrt{2}$. We consider a potential, such that CP is conserved and the flavor-changing neutral currents (FCNC) are suppressed.

The new Higgs sector Lagrangian is

$$\mathscr{L}_{Higgs} = \mathscr{L}_{Kin} + \mathscr{L}_Y - V$$

but now

$$\mathscr{L}_{kin} = (D_\mu\Phi_1)^\dagger D^\mu\Phi_1 + (D_\mu\Phi_2)^\dagger D^\mu\Phi_2$$

$$M_W = \frac{1}{2}\nu g,\ M_Z = \frac{M_W}{\cos\theta_W},\quad M_A = 0,$$

here and in the following ν is equal $\nu = \sqrt{\nu_1^2 + \nu_2^2}$ and

$$\mathscr{L}_Y = \sum_{i.j}\left(y_{ij}^{(u)}\overline{\Psi}_{Li}\Phi_1^c u_{Rj} + y_{ij}^{(d)}\overline{\Psi}_{Li}\Phi_2 d_{Rj}\right),$$

with

$$\Phi_1 = \begin{pmatrix}\phi_1 + i\phi_2\\ \phi_3 + i\phi_4\end{pmatrix},\quad \Phi_2 = \begin{pmatrix}\phi_5 + i\phi_6\\ \phi_7 + i\phi_8\end{pmatrix},\quad \langle\phi_3\rangle_0 = \nu_1,\quad \langle\phi_7\rangle_0 = \nu_2,$$

$$\langle\phi_1\rangle = \langle\phi_2\rangle = \langle\phi_4\rangle = 0,\quad \langle\phi_5\rangle = \langle\phi_6\rangle = \langle\phi_8\rangle = 0,$$

$$M_t = \frac{1}{\sqrt{2}}Y_t\nu_1,\quad M_b = \frac{1}{\sqrt{2}}Y_b\nu_2,$$

where $y_{ij}^{(u,d)}$ are the Yukawa couplings [13], the upper-indexes (u,d) are related with the up and down quark sectors. (L,R) refer to the left quark doublet and right quark singlet. The potential is [14]

$$
\begin{aligned}
V =& \mu_1^2 \Phi_1^\dagger \Phi_1 + \mu_2^2 \Phi_2^\dagger \Phi_2 + \lambda_1 \left(\Phi_1^\dagger \Phi_1\right)^2 + \lambda_2 \left(\Phi_2^\dagger \Phi_2\right)^2 + \lambda_3 \left(\Phi_1^\dagger \Phi_1\right)\left(\Phi_2^\dagger \Phi_2\right) \\
&+ \lambda_4 \left(\Phi_1^\dagger \Phi_2\right)\left(\Phi_2^\dagger \Phi_1\right) + \frac{1}{2}\lambda_5 \left[\left(\Phi_1^\dagger \Phi_2\right)^2 + \left(\Phi_2^\dagger \Phi_1\right)^2\right] + V_{ad} \,,
\end{aligned}
$$

$$
V_{ad} = -\left(m_{12}^2 \Phi_1^\dagger \Phi_2 + \mathrm{h.c.}\right) + \left\{\left[\lambda_6 \left(\Phi_1^\dagger \Phi_1\right) + \lambda_7 \left(\Phi_2^\dagger \Phi_2\right)\right]\left(\Phi_1^\dagger \Phi_2\right) + \mathrm{h.c.}\right\}
$$

We will consider the simplest case in which the additional potential does not contribute, i.e., $V_{ad} = 0$ due to CP conservation and the absence of neutral currents that change flavour [15], [16]. The following equations for the VeVs are obtained using the Stability Condition

$$
\mu_1^2 + \lambda_1 v_1^2 + \Lambda v_2^2 = 0, \quad \mu_2^2 + \lambda_2 v_2^2 + \Lambda v_1^2 = 0\,, \quad \Lambda \equiv \frac{1}{2}\left[\lambda_3 + \lambda_4 + \lambda_5\right],
$$

$$
v_2^2 = \frac{\mu_1^2 \Lambda - \lambda_1 \mu_2^2}{\lambda_1 \lambda_2 - \Lambda^2}, \qquad v_1^2 = \frac{\Lambda \mu_2^2 - \mu_1^2 \lambda_2}{\lambda_1 \lambda_2 - \Lambda^2}.
$$

From the mass matrix, which is defined as

$$
M_{ij}^2 = \frac{1}{2} \left.\frac{\partial^2 V}{\partial \phi_i \partial \phi_j}\right|_{\phi_3 = v_1/\sqrt{2}, \phi_7 = v_2/\sqrt{2}}, \quad i,j = 1, \ldots 8,
$$

and by means of the transformations

$$
U_1 \begin{pmatrix} \phi_1^+ \\ \phi_2^+ \end{pmatrix} = \begin{pmatrix} G^+ \\ H^+ \end{pmatrix}, \quad U_2 \begin{pmatrix} h_1 \\ h_2 \end{pmatrix} = \begin{pmatrix} H^0 \\ h^0 \end{pmatrix}, \quad U_3 \begin{pmatrix} g_1 \\ g_2 \end{pmatrix} = \begin{pmatrix} G^0 \\ A^0 \end{pmatrix}
$$

one obtains the equations for the masses and the eigenstates of the physical particles [17]

$$
\phi_3 = \frac{v_1 + h_1}{\sqrt{2}}, \quad \phi_4 = \frac{g_1}{\sqrt{2}}, \quad \phi_7 = \frac{v_2 + h_2}{\sqrt{2}}, \quad \phi_8 = \frac{g_2}{\sqrt{2}},
$$

$$
M_{G^\pm}^2 = 0, \quad M_{H^\pm}^2 = -\frac{1}{2}\left(\lambda_4 + \lambda_5\right) v^2, \quad M_{A^0}^2 = -\lambda_5 v^2,
$$

$$
M_{H^0,h^0}^2 = \left[\lambda_1 v_1^2 + \lambda_2 v_2^2 \pm \sqrt{\left(\lambda_1 v_1^2 - \lambda_2 v_2^2\right)^2 + \left(2 v_1 v_2 \Lambda\right)^2}\right],
$$

$$
\Lambda \equiv \frac{1}{2}\left[\lambda_3 + \lambda_4 + \lambda_5\right]
$$

Bounds on the quartic Higgs coupling parameters

The positive values of the masses of the scalar particles and the stability condition impose the following restrictions on the quartic couplings, that should be taken into account in the analysis of the energy dependence of the mass spectrum of the scalar particles.

$$\lambda_1 \geq 0, \quad \lambda_2 > 0, \quad \lambda_3 \geq -2\sqrt{\lambda_1\lambda_2}, \quad \lambda_4 \leq |\lambda_5|, \quad \lambda_5 \leq 0,$$
$$(\lambda_3+\lambda_4 \pm \lambda_5) \geq -2\sqrt{\lambda_1\lambda_2},$$
$$(\lambda_3+\lambda_4) \leq 0, \quad 4\lambda_2\lambda_1 - (\lambda_3+\lambda_4 \pm \lambda_5) > 0.$$

To obtain the former result that are in agreement with [19] we determined the minimum using the method of Lagrange multipliers for problems with inequality constraints.

RGE IN THE 2HDM, RESULTS AND CONCLUSIONS

The evolution of the Yukawa Couplings in the Two Higgs Doublet Model was obtained in [18]. The one loop RGE for the gauge couplings, Yukawa couplings and the VeV's in the case where $y_u \simeq y_d$ are

$$\frac{dg_i(t)}{dt} = \frac{1}{(4\pi)^2} b_i g_i^3(t), \quad b_i = (\frac{21}{5}, -3, -7),$$
$$\frac{dy_{u,d}}{dt} = \frac{1}{(4\pi)^2}\left\{-f_{u,d}(t) + \frac{9}{2}y_{u,d}^2 + \frac{1}{2}y_{d,u}^2\right\} y_{u,d},$$
$$\frac{dv_{u,d}}{dt} = \frac{1}{(4\pi)^2}[f_v(t) - 3y_{u,d}^2]v_{u,d},$$

where

$$f_{u,d,v} = \sum C_i^{u,d,v} g_i^2,$$

and

$$f_u(t) = (\frac{17}{20}g_1^2 + \frac{9}{4}g_2^2 + 8g_3^2), \quad f_d(t) = (\frac{1}{4}g_1^2 + \frac{9}{4}g_2^2 + 8g_3^2), \quad f_v(t) = (\frac{9}{20}g_1^2 + \frac{9}{4}g_2^2).$$

The solutions for the gauge couplings, Yukawa couplings and the VeV's are

$$g_l(t) = \frac{g_l(t_0)}{\sqrt{1 - \frac{2b_l g_l^2(t_0)(t-t_0)}{(4\pi)^2}}},$$
$$y_{u,d}^2(t) = \frac{y_{u,d}^2(t_0)}{r_{u,d}(t_0)} \frac{r_{u,d}(t)}{\left(1 - \frac{w_{u,d}(0)}{(4\pi)^2}\int_{t_0}^t \left(9r_{u,d}(t) + \frac{w_{d,u}(t)}{w_{u,d}(t)} r_{d,u}(t)\right) dt\right)},$$
$$\frac{v_{u,d}(t)}{v_{u,d}(t_0)} = \sqrt{r_v(t)}\,(h_{u,d}(t))^{-3}, \quad h_{u,d}(t) \equiv \frac{1}{(4\pi)^2}\exp\int_{t_0}^t y_{u,d}^2\, dt,$$

where $r_{u,d,v}(t)$ are

$$r_{u,d}(t) = \exp\frac{-2}{(4\pi)^2}\int f_{u,d}(t)\,dt = \Pi_{k=1}^{k=3}\left[\frac{g_k^2(t_0)}{g_k^2(t)}\right]^{\frac{C_k^{u,d}}{b_k}},$$

$$r_v(t) = \exp\frac{2}{(4\pi)^2}\int_{t_0}^{t} f_v(t)\,d\tau = \prod_{k=1}^{k=2}\left[\frac{g_k^2(t)}{g_k^2(t_0)}\right]^{\frac{C^v}{b_k}},$$

and

$$\frac{\tan\beta(t)}{\tan\beta(t_0)} \equiv \frac{v_u(t)/v_u(t_0)}{v_d(t)/v_d(t_0)} = \left(\frac{h_d(t)}{h_u(t)}\right)^3.$$

The evolution equations for the quartic couplings $\lambda_1,\ldots,\lambda_5$ are the following

$$\begin{aligned}
(4\pi)^2\frac{d\lambda_1}{dt} =& 24(\lambda_1)^2 - 3\lambda_1\left[3g^2+(g')^2-4y_u^2\right]+2(\lambda_3)^2+(\lambda_4)^2+(\lambda_5)^2,\\
&+2\lambda_3\lambda_4 + 12(\lambda_6)^2+\frac{3}{8}\left(g^2+(g')^2\right)^2+\frac{3}{4}g^4-6y_u^4,\\
(4\pi)^2\frac{d\lambda_2}{dt} =& \Big\{24(\lambda_2)^2-3\lambda_2\left[3g^2+(g')^2-4y_d^2\right]+2(\lambda_3)^2+(\lambda_4)^2+(\lambda_5)^2\\
&+2\lambda_3\lambda_4+12(\lambda 7)^2+\frac{3}{8}\left[(g')^2+g^2\right]^2+\frac{3}{4}g^4-6y_d^4\Big\},\\
(4\pi)^2\frac{d\lambda_3}{dt} =& 4(\lambda_3)^2+4(3\lambda_3+\lambda_4)(\lambda_1+\lambda_2)-3\lambda_3\left[3g^2+(g')^2-2\left(y_t^2+y_d^2\right)\right]\\
&+2(\lambda_4)^2+2(\lambda_5)^2+\frac{3}{4}\left[g^2-(g')^2\right]^2\\
&+\frac{3}{2}g^4-12y_t^2y_d^2+4(\lambda_6)^2+4(\lambda_7)^2+16\lambda_6\lambda_7,\\
(4\pi)^2\frac{d\lambda_4}{dt} =& 4(\lambda_4)^2+4\lambda_4(\lambda_1+\lambda_2+2\lambda_3)-3\lambda_4\left[3g^2+(g')^2-2\left(y_t^2+y_d^2\right)\right]\\
&+8(\lambda_5)^2+3g^2(g')^2+12y_t^2y_d^2+10(\lambda_6)^2+10(\lambda_7)^2+4\lambda_6\lambda_7,\\
(4\pi)^2\frac{d\lambda_5}{dt} =& \lambda_5\Big\{4(\lambda_1+\lambda_2+2\lambda_3+3\lambda_4)-3\left[3g^2+(g')^2-2\left(y_t^2+y_d^2\right)\right]\\
&+10(\lambda_6)^2+10(\lambda_7)^2+4\lambda_6\lambda_7\Big\}.
\end{aligned}$$

The number of parameters of the scalar potential of the 2HDM model is large and cannot be determined from experimental data. To be able to perform phenomenological analysis of the model one has to make some assumptions. We choose the universality assumption for the parameters of the quartic potential

$$\lambda_1(E_{GU}) = \lambda_2(E_{GU}) = \lambda_3(E_{GU}) = \lambda_4(E_{GU}) = |\lambda_5(E_{GU})| \tag{4}$$

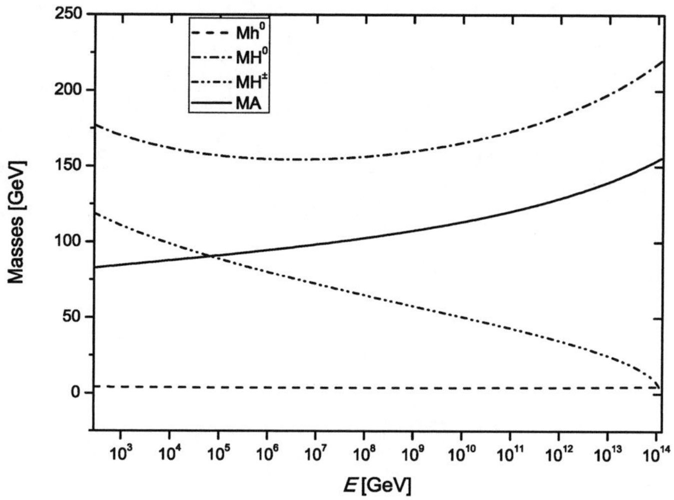

FIGURE 4. The evolution of the Higgs boson masses in the 2HDM. Notice that $M_{H^\pm}(E_{GU}) = 0$ follows from the universality of the potential parameters λ_i.

and for the Yukawa *top* and *bottom* quark couplings

$$Y_t(E_{GU}) = Y_b(E_{GU}) = 0.6.$$

For the remaining parameters we have used the following values

$$E_{GU} \sim 10^{14}\,\text{GeV}, \quad m_t = 174.1$$
$$(g_1, g_2, g_3)_{m_t} = (0,447, 0.626, 1.164)$$
$$(g_1 = g_2)_{GU} = 0.529.$$

Using these assumptions we have obtained the results that are shown in Figs. 4, 5 and 6. From these figures one can observe the following general pattern of the renormalization group evolution of the parameters of the 2HDM:

1. The evolution of the 5 Higgs masses shown in Fig. 4 is rather slow and the mass of the lightest neutral Higgs boson is almost constant. The mass of the charged Higgs boson is decreasing monotonically with energy.
2. One can see that there is no general pattern of the evolution of λ_i's with the exception of λ_1 and λ_2 which run almost identically. Each of the remaining λ_i's evolve differently. Notice that the constraints on the values of λ_i are fulfilled in the whole range $[m_t, E_{GU}]$.
3. From Fig. 6 one can see that the evolution of Y_t and Y_b is almost identical. This means that the ratio of the *top* and *bottom* quark masses remains constant. The quark mass hierarchy thus cannot be explained through the renormalization group evolution.

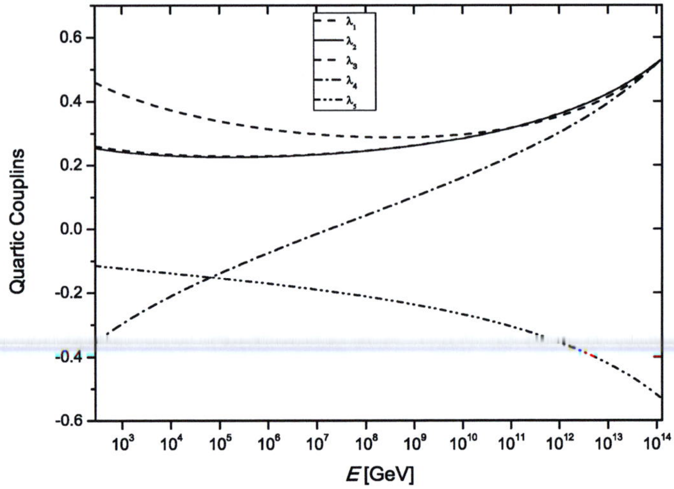

FIGURE 5. The evolution of the Higgs scalar potential parameters λ_i in the 2HDM.

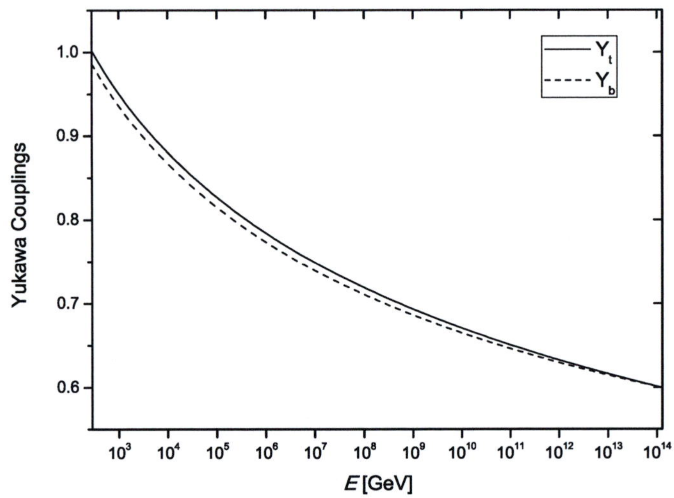

FIGURE 6. The evolution of the Yukawa couplings Y_t and Y_b in the 2HDM.

In Table 1 we show the initial and final values of the RGE evolution of the quartic couplings and the Higgs masses in the one loop approximation. This table very well illustrates the pattern of evolution shown in Figs. 4 and 5.

TABLE 1. The evolution of the scalar potential parameters λ_i and of the Higgs masses. The values at E_{GU} were assumed and the values at m_t were numerically derived from the renormalization group equations.

	λ_1	λ_2	λ_3	λ_4	λ_5	M_{h^0}	M_{H^0}	$M_{H^\pm}$	M_A
E_{GU}	0.529	0.529	0.529	0.529	−0.529	4.63	220.2	0.0	155.75
m_t	0.260	0.25	0.46	−0.35	−0.11	4.19	177.2	118.8	82.89

CONCLUSIONS

It is a common belief that the Standard Model and its extensions are only the effective theories and that there should appear new degrees of freedom or new effects at higher energies that are presently not accessible to the direct experimental analysis. However the low energy values of the observables can give information at what energy these new effects should appear and the renormalization group is a tool of choice for the search of the value of the energy at which the new physics should appear.

The most sensible observable for a search of new physics is the mass of the Higgs boson. Our analysis does not give a hint what mass the Higgs particle might have, but from the mass of the Higgs boson it predicts at what energy the Standard Model or its two Higgs extension become unstable or singular and no longer correctly describe our world.

We have analyzed the renormalization group evolution of various observables in the Standard Model and in the two Higgs doublet extension of the Standard Model. The main physical conditions imposed on the theory are the stability and the triviality of the underlying gauge field theory which is expressed by the high energy behavior of the quartic coupling constants of the scalar potential.

Our results can be divided into two categories: first one for the Standard Model and the second one for the two Higgs extension of the Standard Model.

The Standard model analysis is relatively simple, because there is only one quartic coupling constant λ_H and only one Higgs boson, whose mass M_H is directly related to the quartic coupling constant. We explicitly solve the RGE for λ_H and show in Fig. 1 what is the allowed range of allowed values of λ_H for a given energy. Fig. 3 is the counterpart of Fig. 1 and it shows the allowed range of the Higgs mass for a given energy. Fig. 3 is the most important result concerning the Standard Model, because it tells at what energy the new physics should appear if the Higgs mass is known. The interesting conclusion that can be seen from Fig. 3 is that for the Standard Model Higgs boson mass is in the range

$$148.5 \leq M_H \leq 193.1$$

there is no need of new physics up to the Planck mass and the Standard Model may not need any modification.

The analysis of the two Higgs extension of the Standard Model is more involved. There are at least five quartic coupling constants and five Higgs bosons and the most general analysis cannot be very conclusive. For this reason we have used the additional assumption of the universality, Eq. (4), which gives the same weight to various components of the scalar Higgs potential at the Planck energy E_{GU}. With this assumption we can also predict the five Higgs masses and their evolution. The results for this model are contained in Figs. 4, 5, 6 and demonstrate the following facts:

1. The universality assumption gives a theory valid in the whole range of energies from the top quark mass m_t to the Planck energy E_{GU}.
2. The variation of the Higgs masses is slow.
3. The evolution of the *top* and *bottom* Yukawa couplings is almost identical, which means that the *top-bottom* quark mass hierarchy is preserved throughout the whole energy range.

Finally one can also see that if the neutral Higgs has the mass ~ 180 GeV then it is compatible with both Standard Model and its two Higgs boson extension and both theories do not support the appearance of new physics up to the Planck scale.

ACKNOWLEDGMENTS

We would like to thank Professor Miguel Angel Pérez Angón for inviting us to participate in the symposium celebrated in honor of Augusto García.

This work was partially supported by COFAA, EDI from IPN, CONACyT and SNI-Mexico.

REFERENCES

1. P.W. Higgs, Phys. Lett. **12** (1964) 132; Phys. Rev. Lett.**13** (1964) 508; Phys. Rev. **145** (1966) 1156.
2. M. Carena, H.E. Haber, Prog. Part. Nucl. Phys. **50**(2003) 63-152; http://public.web.cern.ch/public/en/Research/Research-en.html
3. J.F. Gunion, H.E. Haber, G.L. Kane and S. Dawson, *The Higgs Hunter´s Guide* Addison-Wesley, Reading, MA, 1990).
4. S. Glashow, Nucl. Phys. **22** (1961) 579; S. Weinberg, Phys. Rev.Lett.**19**, 1264 (1967); A. Salam, *Weak and electromagnetic interactions* in *Elementary Particle Theory*, p. 367, W. Svartholm, editor, Stockholm, Almquist and Wiksell, 1968.
5. H. Arason, et al, Phys. Rev. D **46**, 3945 (1992); C.R. Das, M.K. Parida, Eur. Phys. J. C **20**, 121-132 (2001)
6. The Review of Particle Physics. C. Amsler et al. (Particle Data Group), Physics Letters B **667**, 1 (2008)
7. P. Kielanowski, S.R. Juárez W. and J.H. Montes de Oca Y. Phys.Rev.D**78**, 116010(8) 2008.
8. M. Luo and Y. Xiao, Phys. Rev. Lett.**90**, 011601 (2003); M. Luo and Y. Xiao, Phys. Rev. Lett.**90**, 011601 (2003); M Luo, H. Wang and Y. Xiao, *Two Loop Renormalization Group Equations in General Gauge Field Theories*, hep-ph/0211440.
9. P. Kielanowski, S.R. Juárez W. and H.G. Solís-Rodríguez, Phys. Rev. D **72**, 096003 (2005) (and references therein); H.G. Solís R., S.R. Juárez W. and P. Kielanowski, AIP CP **857**:202-209, 2006. P. Kielanowski, S.R. Juárez W.,AIP CP **960** 85-89 (2007)
10. P. Kielanowski, S.R. Juárez W. and G. Mora, Phys. Lett. B**479** (1-3), 181-189 (2000).
11. S.R. Juárez W., S.F. Herrera H., P. Kielanowski and G. Mora; Phys.Rev.D **66**(116007) 1-11 (2002).
12. Riccati Boyce, W. E. and DiPrima, R. C. *Elementary Differential Equations and Boundary Value Problems*, 4th ed. New York: Wiley, 1986, p. 87; Einar Hille, Lectures on Ordinary Differential Equations. Addison-Wesley Pub. Co.(1969).
13. K. Inoue, A. Kakuto and Y. Nakano, Prog. Theor. Phys. **63** (1980) 234.

14. Y-L. Wu and L. Wolfenstein Phys. Rev. Lett. **73**, 1762 (1994)
15. P.M. Ferreira and D. R. T. Jones, arXiv:hep-ph/0903.2856.
16. S. Glashow, S. Weinberg, Phys. Rev D **15**, 1958 (1977); L.J. Hall and S. Weinberg, Phys. Rev D **48**, R979 (1993)
17. S.R. Juárez W. and D. Morales C., AIP Conf. Proc. **1026** p. 294-296 (2008).
18. J.H. Montes de Oca Y., S.R. Juárez W., P. Kielanowski, AIP Conf. Proc. **1026** p. 291-293 (2008).
19. N.G. Deshpande and E. Ma, Phys. Rev. D **18**, 2574 (1978); K.S. Babu and B. Ma, ibid. **31**, 2861 (1985); **33**, 347 (E) (1986); D. Kominis and R. Sekhar, Phys. Lett. B **304** (1993) 152; J.F. Gunion, H.E. Haber Phys. Rev. D **67**, 075019 (2003); M. Hashimoto, Sh. Kanemura, Phys. Rev. D **70**, 055006 (2004); P.M. Ferreira, R. Santos, A. Barroso, Physics Letters B **603** (2004) 219-229.
20. S. Kraml, (164 additional authors not shown) arXiv:hep-ph/0608079 CERN-CPNSH Workshop, May 2004 - Dec. 2005. Ed. by Sabine Kraml, et. al. (GENEVA 2006).

Neutral currents production in LHC for 331 models

R. Martinez and F. Ochoa

Departamento de Física, Universidad Nacional, Bogotá-Colombia

Abstract. A brief review about the production of a Z' resonance in the framework of the 331 models is given. Their signatures at CERN LHC is highlighted by studying the Z' production and decay features, including different final states and one-loop corrections..

Keywords: 331 models, Z' physics, LHC
PACS: 14.70.-e, 11.10.Gh, 11.15.-q, 12.60.Cn

INTRODUCTION

In many extensions of the Standard Model (SM), new massive and neutral gauge bosons, called Z', are predicted [1]. The detection of a Z' resonance has became in a matter of high priority in particle physics, which could reveal many features about the underlying unified theory. Indirect search for this neutral boson has been carried out at LEP, through the mixing with the Z boson [2]. Direct Z' production has been searched at the Tevatron [3, 4]. The discovery potential for a Z' particle has been explored at the forthcoming Large Hadron Collider (LHC) [5] in the $M_{Z'} \approx 1-5$ TeV range. The search for this particle has also been explored in the planned International Liner Collider (ILC) [6]. If a Z' signal is detected, further analyses will be neccesary in order to determine all its properties which will allow to test the compatibility of the theoretical models with the experimental data. There are many theoretical models which predict a Z' mass in the TeV level, where the most popular are the E_6 motivated models [1], the Left-Right Symetric Model (LRM) [7], the Z' in Little Higgs scenario [8] and the Sequential Standard Model (SSM), which has heavier couplings than those of the SM Z boson. On the other hand, the models with gauge symmetry $SU(3)_c \otimes SU(3)_L \otimes U(1)_X$, also called 331 models [9, 10], arise as an interesting alternative with Z' boson and many well-established motivations. First of all, from the cancellation of chiral anomalies [11] and asymptotic freedom in QCD, the 3-3-1 models can explain why there are three fermion families. Secondly, since the third family is treated under a different representation, the large mass difference between the heaviest quark family and the two lighter ones may be understood [12]. Thirdly, the models have a scalar content similar to the two Higgs doublet model (2HDM), which allow to predict the quantization of electric charge and the vectorial character of the electromagnetic interactions [13]. Also, these models contain a natural Peccei-Quinn symmetry, necessary to solve the strong-CP problem [14].

In this talk we present some signatures of the 331 models including fermion and higgs events as final state at CERN LHC. We also consider the one-loop effects due to new heavy J quarks and $K^{Q_{1,2}}$ gauge bosons predicted by the model. The analyses discussed

CP1259, *First CINVESTAV-UNAM Symposium on High Energy Physics*
edited by J. C. D'Olivo, R. López-Fernández, and M. Á. Pérez

TABLE 1. Fermionic content for three generations in the FLT model, where $j = 1,2,3$ for each lepton generation

Representation	Q_ψ	X_ψ
$q_{1,2L} = \begin{pmatrix} d,s \\ u,c \\ J_{1,2} \end{pmatrix}_L : \mathbf{3}^*$	$\begin{pmatrix} -\frac{1}{3} \\ \frac{2}{3} \\ -\frac{1}{3} \end{pmatrix}$	$X^L_{q_{1,2}} = 0$
$(d,s)_R;\ (u,c)_R;\ J_{1,2R} : \mathbf{1}$	$-\frac{1}{3};\ \frac{2}{3};\ -\frac{1}{3}$	$X^R_{q_{1,2}} = -\frac{1}{3}, \frac{2}{3}, -\frac{1}{3}$
$q_{3L} = \begin{pmatrix} t \\ b \\ J_3 \end{pmatrix}_L : \mathbf{3}$	$\begin{pmatrix} \frac{2}{3} \\ -\frac{1}{3} \\ \frac{2}{3} \end{pmatrix}$	$X^L_{q_3} = \frac{1}{3}$
$t_R;\ b_R;\ J_{3R} : \mathbf{1}$	$\frac{2}{3};\ -\frac{1}{3};\ \frac{2}{3}$	$X^R_{q_3} = \frac{2}{3}, -\frac{1}{3}, \frac{2}{3}$
$\ell_{jL} = \begin{pmatrix} \nu_j \\ e_j \\ N_j \end{pmatrix}_L : \mathbf{3}$	$\begin{pmatrix} 0 \\ -1 \\ 0 \end{pmatrix}$	$X^L_{\ell_j} = -\frac{1}{3}$
$e_{jR};\ N_{jR}$	$-1;\ 0$	$X^R_{\ell_j} = -1,\ 0$

here and further details about Z' signatures in 331 models, can be found in references [15, 16, 17].

THE 331 SPECTRUM

The fermionic structure is shown in Tab. 1 where all leptons transforms as $(\mathbf{3}, \mathbf{X}^L_\ell)$ and $(\mathbf{1}, \mathbf{X}^R_\ell)$ under the $(SU(3)_L, U(1)_X)$ sector, with $\mathbf{X}^L_\ell$ and $(\mathbf{X}^R_\ell)$ the $U(1)_X$ generators associated with the left- and right-handed leptons, respectively; while the quarks transforms as $(\mathbf{3}^*, \mathbf{X}^L_{q_{1,2}})$, $(\mathbf{1}, \mathbf{X}^R_{q_{1,2}})$ for the first two families, and $(\mathbf{3}, \mathbf{X}^L_{q_3})$, $(\mathbf{1}, \mathbf{X}^R_{q_3})$ for the third family, each one with its $U(1)_X$ values for the left- and right-handed quarks. The quantum numbers $\mathbf{X}_\psi$ for each representation are given in the third column from Tab. 1, where the electric charge is defined by

$$Q = T_3 + \beta T_8 + XI, \tag{1}$$

with $T_3 = 1/2\,\mathrm{diag}(1,-1,0)$, $T_8 = (1/2\sqrt{3})\,\mathrm{diag}(1,1,-2)$ and $\beta = -1/\sqrt{3}$ corresponding to the Foot-Long-Truan (FLT) model [10].

For the scalar sector, the following triplet fields are introduced

$$\chi = \begin{pmatrix} \chi_1^0 \\ \chi_2^\mp \\ \frac{1}{\sqrt{2}}(\xi_\chi + \nu_\chi + i\zeta_\chi) \end{pmatrix}; \rho = \begin{pmatrix} \rho_1^\pm \\ \frac{1}{\sqrt{2}}(\xi_\rho + \nu_\rho + i\zeta_\rho) \\ \rho_3^\pm \end{pmatrix}; \eta = \begin{pmatrix} \frac{1}{\sqrt{2}}(\xi_\eta + \nu_\eta + i\zeta_\eta) \\ \eta_2^\mp \\ \eta_3^0 \end{pmatrix},$$

TABLE 2. Higgs boson spectrum in mass eigenstates. f and λ_i are coupling constants of the Higgs potential and H_1^0, $\overline{H_1^0}$ are complex fields

Higgs Boson	Squared Mass
$h^0 = S_{\beta_H}\xi_\rho + C_{\beta_H}\xi_\eta$	$M_{h^0}^2 \simeq \frac{4}{v_\eta^2+v_\rho^2}\left[\lambda_2 v_\rho^4 + 2\lambda_6 v_\rho^2 v_\eta^2 + \lambda_3 v_\eta^4\right]$
$H^0 = -C_{\beta_H}\xi_\rho + S_{\beta_H}\xi_\eta$	$M_{H^0}^2 \simeq -\sqrt{2} f v_\chi \left(\frac{v_\eta}{v_\rho} + \frac{v_\rho}{v_\eta}\right)$
$H_3^0 \simeq \frac{1}{\sqrt{2}}\xi_\chi$	$M_{H_3^0}^2 \simeq 4\lambda_1 v_\chi^2$
$H_1^0(\overline{H_1^0}) = \eta_3^0(\eta_3^{0*})$	$M_{H_1^0}^2 \simeq \frac{1}{2}\lambda_7 v_\chi^2 - \frac{1}{\sqrt{2}} f v_\chi \frac{v_\rho}{v_\eta}$
$H^\pm = C_{\beta_H}\rho_1^\pm + S_{\beta_H}\eta_2^\pm$	$M_{H^\pm}^2 \simeq -\frac{1}{\sqrt{2}} f v_\chi \left(\frac{v_\eta}{v_\rho} + \frac{v_\rho}{v_\eta}\right)$
$H_2^\pm = \rho_3^\pm$	$M_{H_2^\pm}^2 \simeq \frac{1}{2}\lambda_8 v_\chi^2 - \frac{1}{\sqrt{2}} f v_\chi \frac{v_\eta}{v_\rho}$

with vacuum expectation value (VEV) $\langle\chi\rangle^T = (0,0,v_\chi)$, which induces the masses to the third fermionic components, and $\langle\rho\rangle^T = (0,v_\rho,0)$ and $\langle\eta\rangle^T = (v_\eta,0,0)$, which gives masses to the quarks of type up and down respectively [18]. The mass eigenstates of the Higgs bosons are written in the Tab. 2 [19], where the rotation angle β_H is defined by[18]

$$t_{\beta_H} = \tan\beta_H = \frac{v_\rho}{v_\eta}, \tag{2}$$

In the gauge boson spectrum, it is associated with the group $SU(3)_L$ which transform according to the adjoint representation and are written in the form [19]

$$\mathbf{W}_\mu = W_\mu^\alpha G_\alpha = \frac{1}{2}\begin{bmatrix} W_\mu^3 + \frac{1}{\sqrt{3}}W_\mu^8 & \sqrt{2}W_\mu^+ & \sqrt{2}K_\mu^{Q_1} \\ \sqrt{2}W_\mu^- & -W_\mu^3 + \frac{1}{\sqrt{3}}W_\mu^8 & \sqrt{2}K_\mu^{Q_2} \\ \sqrt{2}K_\mu^{-Q_1} & \sqrt{2}K_\mu^{-Q_2} & -\frac{2}{\sqrt{3}}W_\mu^8 \end{bmatrix}, \tag{3}$$

where $Q_{1,2} = (1/2)(\pm 1 + \sqrt{3}\beta)$, with $\beta = -1/\sqrt{3}$. As for the gauge field associated with $U(1)_X$, it is represented as $\mathbf{B}_\mu = B_\mu \mathbf{I}_{3\times 3}$ which is a singlet under $SU(3)_L$, and has no electric charge. For the charged sector, the following mass eigenvalues are obtained at tree-level

$$M_{W^\pm}^2 = \frac{g_L^2}{4}\left(v_\rho^2 + v_\eta^2\right), \quad M_{K^{\pm Q_1}}^2 = \frac{g_L^2}{4}\left(v_\chi^2 + v_\eta^2\right), \quad M_{K^{\pm Q_2}}^2 = \frac{g_L^2}{4}\left(v_\chi^2 + v_\rho^2\right) \tag{4}$$

The neutral sector is [19]

$$\begin{aligned} A_\mu &= S_W W_\mu^3 - \frac{C_W}{\sqrt{3}}\left(T_W W_\mu^8 - \sqrt{3 - T_W^2}B_\mu\right), \\ Z_\mu &= C_W W_\mu^3 + \frac{S_W}{\sqrt{3}}\left(T_W W_\mu^8 - \sqrt{3 - T_W^2}B_\mu\right), \\ Z'_\mu &= -\frac{1}{\sqrt{3}}\left(\sqrt{3 - T_W^2}W_\mu^8 + T_W B_\mu\right), \end{aligned} \tag{5}$$

where the Weinberg angle is defined as $S_W = \sqrt{3}g_X/\sqrt{3g_L^2+4g_X^2}$, with g_L and g_X the coupling constants of the groups $SU(3)_L$ and $U(1)_X$ respectively. The corresponding eigenvalues at tree-level are

$$M_{A_\mu}^2 = 0\,;\; M_{Z_\mu}^2 \simeq \frac{g_L^2}{4C_W^2}\left(v_\rho^2+v_\eta^2\right);\; M_{Z'_\mu}^2 \simeq \frac{g_X^2}{3T_W^2}v_\chi^2. \tag{6}$$

THE ELECTROWEAK LAGRANGIAN

The couplings in weak eigenstates at tree level are [20]

$$\begin{aligned} \mathcal{L}_D &= i\overline{f}\partial\!\!\!/ f + eQ_f\overline{f}A\!\!\!/ f + \frac{g_L}{2C_W}\left[\overline{f}\gamma_\mu\left(g_v^f - g_a^f\gamma_5\right)fZ^\mu + \overline{f}\gamma_\mu\left(\tilde{g}_v^f - \tilde{g}_a^f\gamma_5\right)fZ^{\mu\prime}\right] \\ &+ \frac{g_L}{\sqrt{2}}\overline{U_L}W\!\!\!\!/^{\,+}D_L + \frac{g_L}{\sqrt{2}}\overline{U_L}\mathbf{K}_{UJ}^{+}\mathbf{J}_L + \frac{g_L}{\sqrt{2}}\overline{D_L}\mathbf{K}_{DJ}^{+}J_L + \text{h.c.}, \end{aligned} \tag{7}$$

with $\mathbf{K}_{UJ}^{\pm} = \text{diag}(K\!\!\!/^{\pm Q_1}, K\!\!\!/^{\mp Q_2}, K\!\!\!/^{\mp Q_2})$, $\mathbf{K}_{DJ}^{\pm} = \text{diag}(K\!\!\!/^{\pm Q_2}, -K\!\!\!/^{\mp Q_1}, -K\!\!\!/^{\mp Q_1})$, f any fermion flavor, $U = (u_1,u_2,u_3)$, $D = (d_1,d_2,d_3)$ and $J = (J_1,J_2,J_3)$ the quark components. The vector and axial-vector couplings g_v^f are the same as the SM couplings for f the SM fermions, while for the new J quarks are $g_v^J = -2\mathbf{Q}_J S_W^2$, and $g_a^J = 0$, with $\mathbf{Q}_J = (Q_{J_1}, Q_{J_2}, Q_{J_3})$ the charges of the J quarks. The Z' couplings are defined as

$$\begin{aligned} \tilde{g}_{v,a}^U &= n\left[\tfrac{1}{\sqrt{3}}\left(\text{diag}(-1,1,1) + \tfrac{\beta T_W^2}{\sqrt{3}}\right) \pm 2Q_U\beta T_W^2\right], \\ \tilde{g}_{v,a}^D &= n\left[\tfrac{1}{\sqrt{3}}\left(\text{diag}(-1,1,1) + \tfrac{\beta T_W^2}{\sqrt{3}}\right) \pm 2Q_D\beta T_W^2\right], \\ \tilde{g}_{v,a}^J &= n\left[\tfrac{2}{\sqrt{3}}\left(\text{diag}(1,-1,-1) + \sqrt{3}\mathbf{Q}_J\beta T_W^2\right) \pm 2\mathbf{Q}_J\beta T_W^2\right], \\ \tilde{g}_{v,a}^N &= n\left[-\tfrac{1}{\sqrt{3}} - \beta T_W^2\right], \\ \tilde{g}_{v,a}^L &= n\left[-\tfrac{1}{\sqrt{3}} - \beta T_W^2 \pm 2Q_L\beta T_W^2\right], \end{aligned} \tag{8}$$

with $n = g_X C_W/2g_L T_W$ and $\beta = -1/\sqrt{3}$.

The cubic interaction for the Higgs$-Z'$ and Higgs$-Z'-Z$ couplings is described by

$$\begin{aligned} i\mathcal{L}^{HZ'} &= \frac{g_L}{2\sqrt{3-T_W^2}}\left[\left(S_{2\beta_H}\right)h^0A^0 + \left(C_{2\beta_H} - T_W^2\right)H^0A^0\right. \\ &\left. + i\left(C_{2\beta_H} + T_W^2\right)H^+H^- + 2iH_1^0\overline{H_1^0} + 2i\left(1-T_W^2\right)H_2^+H_2^-\right](p-q)^\mu Z'_\mu \\ &+ \frac{ig_L M_Z}{\sqrt{3-T_W^2}}\left[\left(S_{2\beta_H}\right)H^0 + \left(C_{2\beta_H} + T_W^2\right)h^0\right]Z^\mu Z'_\mu, \end{aligned} \tag{9}$$

where β_H is the rotation angle defined by Eq. (2).

Z' AT CERN LHC

The design criteria of ATLAS at LHC could reveal a Z' signal at the TeV scale for pp collisions at C.M. energy $\sqrt{s} = 14$ TeV, and integrated luminosity $L = 100 fb^{-1}$. For dielectron final states, it is expected a pseudorapidity below $|\eta| \leq 2.2$, and transverse energy cut $E_T \geq 20$ GeV [5].

The plot in Fig. 1a shows the invariant mass distribution for the dielectron system as final state for a central value $M_{Z'} = 1500$ GeV. The SM Drell-Yan spectrum with the same kinematical conditions is also shown, where the Z' signal exhibit a huge peak above the SM background. The number of events is about 800 signal events/(20 GeV) in the resonance [15]. The total cross section for the same leptonic channel as a function of $M_{Z'}$ is shown in Figr. 1b. For the range 800 GeV $\leq M_{Z'} \leq$2000 GeV, the number of events decreases in the range $5 \times 10^4 \leq N \leq 2400$. Fig. 2a shows the branching ratios $Br = \sigma(pp \to Z' \to f\overline{f})/\sigma_{tot}$ as a function of $M_{Z'}$ for the $u\overline{u}$, $d\overline{d}$, $b\overline{b}$, $t\overline{t}$, e^+e^-,$\nu\overline{\nu}$ and $T\overline{T}$ channels, where T is the J_3 quark which has the same charge as the top quark, as we can see in Tab. 1. A value of $m_T = 300$ GeV is chosen. The heavy top quark T production shows the biggest branching ratio. Fig. 2b exhibit the two body cross sections with $M_{Z'}$ in each fermion channel. In general, the cross section decreases with $M_{Z'}$. It is interesting to see that the quark T exhibit a huge cross section with about $N = 6 \times 10^4$ events at $M_{Z'} = 800$ GeV, and decreases to just 30 events at $M_{Z'} = 5000$ GeV.

Finally, Fig. 3a show the branching ratios for the Higgs channel. As a first aproximation, $\nu_\rho = 10\nu_\eta$, i.e $t_{\beta_H} = 10$ is used. The heavy $H_0^1\overline{H_0^1}$ final state is the most probably higgs decay channel. In the charged sector, there are two kind of heavy higgs $H^\pm$ and $H_2^\pm$, where the later shows a bigger branching ratio than the former. The model exhibit one like-SM higgs boson h^0, which can be produced as two body final state in conjunction with the pseudoscalar A^0 and the weak boson Z. However, this channel have a very

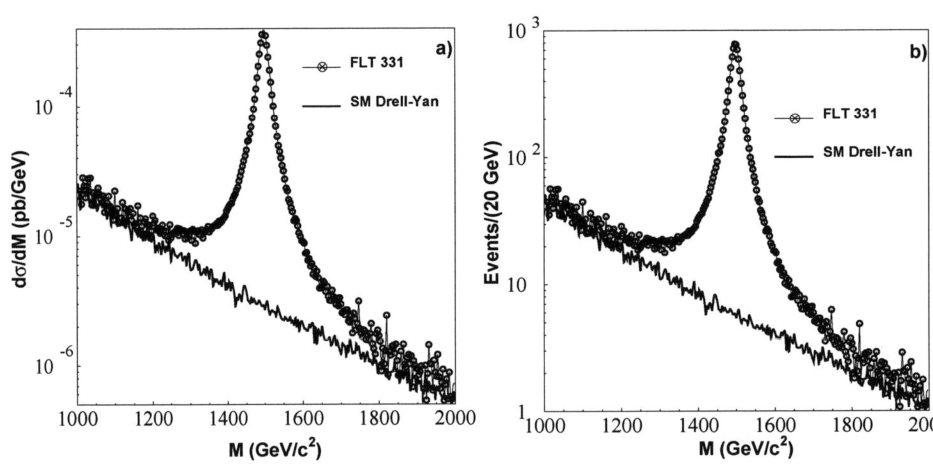

FIGURE 1. a.) Di-electron cross section distribution for $M_{Z'} = 1.5$ TeV at LHC. b.) Total cross section as function of $M_{Z'}$

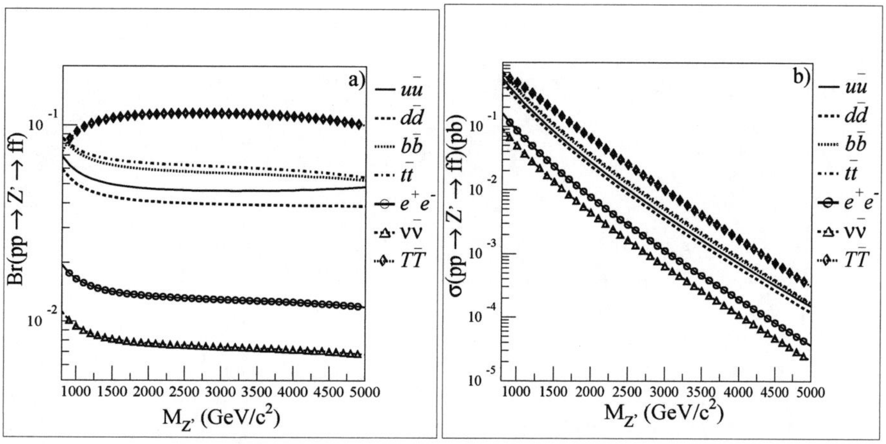

FIGURE 2. a.) Fermion branching ratios as function of $M_{Z'}$ at LHC. b.) The corresponding total cross sections

low branching ratio. Fig. 3b display the total cross sections, where about 1×10^4 events per year is obtained for the $H_0^1\overline{H_0^1}$ final state at $M_{Z'} = 800$ GeV, and just 200 events for the h^0A^0 channel [15].

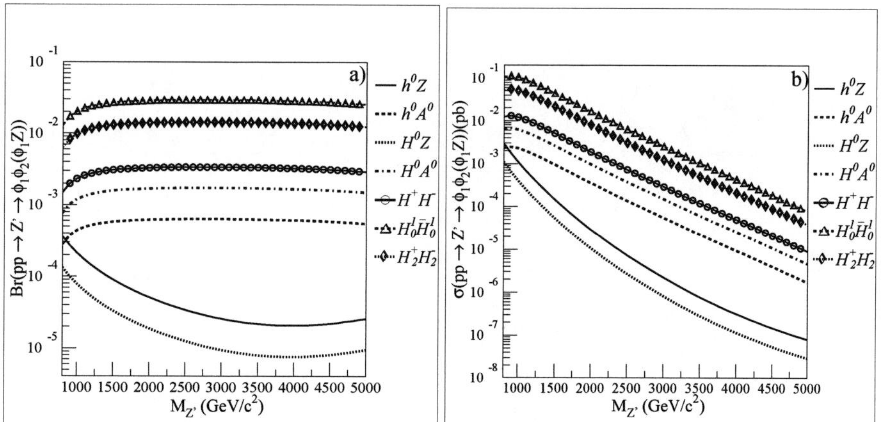

FIGURE 3. a.) Higgs branching ratio as function of $M_{Z'}$ at LHC. b.) The corresponding total cross sections

One-loop correction in Z' decay

The Z' decay width at tree-level is

$$\Gamma_0(Z' \to f\overline{f}) = \frac{g_L^2 M_{Z'} N_c^f}{48\pi C_W^2}\sqrt{1-\mu_f'^2}\left[\left(1+\frac{\mu_f'^2}{2}\right)\left(\widetilde{g}_v^f\right)^2 + \left(1-\mu_f'^2\right)\left(\widetilde{g}_a^f\right)^2\right] R_{QED}R_{QCD}, \quad (10)$$

where $\mu_f'^2 = 4m_f^2/M_{Z'}^2$ takes into account kinematical corrections only important for external heavy fermions. $R_{QED,QCD}$ are global final-state corrections from photons and gluons process. On the other hand, the decay ratio can be modified by one-loop corrections sensitive to heavy particles running into the loop in both the Z' propagator and $Z'f\overline{f}$ vertex as we can see in fig. 4, where contributions from the exotic quarks J from Tab. 1 and the charged bosons $K^{\pm Q_{1,2}}$ are obtained. The oblique corrections corresponds to figs (1a,b,c), and the vertex corrections mediated by the K_μ gauge bosons can be seen in figs. (1d,e,f). Taking a suitable renormalization scheme, the Z' decay width into fermions becomes [16]

$$\Gamma_{Z'\to\overline{f}f} = \frac{g_X^2 M_{Z'} N_c^f}{192\pi T_W^2}\rho_f'\sqrt{1-\mu_f'^2}\left[\left(1+\frac{\mu_f'^2}{2}\right)\left(\widetilde{g}_{v,eff}^f\right)^2 + \left(1-\mu_f'^2\right)\left(\widetilde{g}_{a,eff}^f\right)^2\right] R_{QED}R_{QCD}, \quad (11)$$

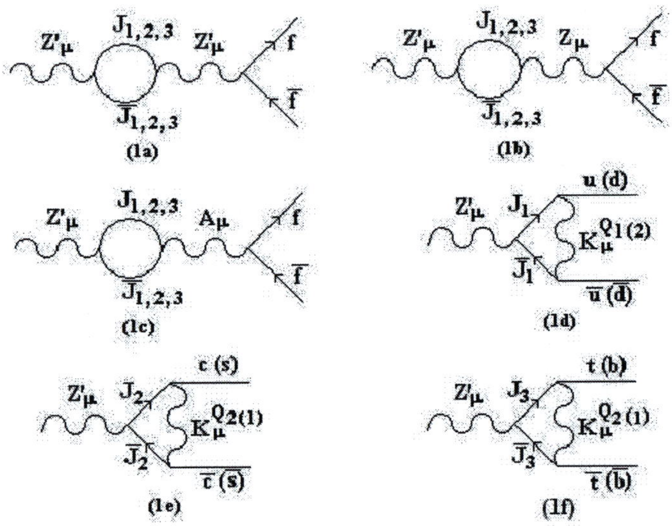

FIGURE 4. One-loop oblique (a,b,c) and vertex (d,e,f) corrections

TABLE 3. Tree-level width, width with oblique corrections, width with oblique plus vertex corrections, correction fraction of the total contributions (δ_f) and with only vertex contributions ($\delta_{q-vertex}$). The data are calculated for $m_J = 2$ TeV

Final states	Γ_{0f}(GeV)	Γ_f^{obl}(GeV)	Γ_f(GeV)	$\delta_f(\%)$	$\delta_{q-vertex}(\%)$
$u\bar{u}(c\bar{c})$	2.293	2.391	2.362	3.02	-1.19
$t\bar{t}$	3.271	3.517	3.481	6.42	-1.04
$d\bar{d}(s\bar{s})$	1.963	1.966	1.938	-1.28	-1.42
$b\bar{b}$	2.974	3.122	3.086	3.76	-1.17
$e^+e^-(\mu^\pm,\tau^\pm)$	0.665	0.722	0.722	8.48	—
$\nu_{e,\mu,\tau}\bar{\nu}_{e,\mu,\tau}$	0.344	0.308	0.308	-10.48	—

where the corrections have a quadratic m_J dependence, which is absorbed into the parameter ρ_f' and the effective couplings given by [16]

$$\begin{aligned}
\widetilde{g}^U_{v,a,eff} &= \tfrac{1}{\sqrt{3}}\left[\mathrm{diag}\left(-1,1,1\right)+\tfrac{\beta\overline{T_W^2}}{\sqrt{3}}\right]\pm 2Q_U\beta\overline{T_W^2},\\
\widetilde{g}^D_{v,a,eff} &= \tfrac{1}{\sqrt{3}}\left[\mathrm{diag}\left(-1,1,1\right)+\tfrac{\beta\overline{T_W^2}}{\sqrt{3}}\right]\pm 2Q_D\beta\overline{T_W^2},\\
\widetilde{g}^N_{v,a,eff} &= -\tfrac{1}{\sqrt{3}}-\beta\overline{T_W^2},\\
\widetilde{g}^L_{v,a,eff} &= -\tfrac{1}{\sqrt{3}}-\beta\overline{T_W^2}\pm 2Q_L\beta\overline{T_W^2},
\end{aligned} \tag{12}$$

where $\rho_f' = 1+\Delta\rho_j$ and $\overline{T_W^2} = T_W^2(1+\varepsilon\Delta\rho_j)$ for $f=\ell$ any lepton, while $\rho_f' = 1+(1/9)\Delta\rho_j$ and $\overline{T_W^2} = T_W^2(1+(\varepsilon+4/9)\Delta\rho_j)$ for $f=q$ any quark. The parameters $\Delta\rho_j$ and ε are

$$\Delta\rho_j \approx 3\left(\frac{g_X}{4\pi T_W}\right)^2 x_J, \qquad \varepsilon = \frac{1-\beta^2 T_W^2}{\beta^2 T_W^2}, \tag{13}$$

with $\beta = -1/\sqrt{3}$ and $x_J = m_J^2/M_{Z'}$. Table 3 shows the values of $\Gamma_{0f},\Gamma_f^{obl},\Gamma_f,\delta_f$ and $\delta_{q-vertex}$ for $m_J = 2$ TeV for each fermion channel, where Γ_{0f} is tree-level width given by Eq. 10, Γ_q^{obl} is the width with only oblique corrections, Γ_f the width including both oblique and vertex corrections, δ_f the deviation due to oblique and vertex corrections defined by $\delta_f = (\Gamma_f-\Gamma_{0f})/\Gamma_{0f}$, and $\delta_{q-vertex}$ the deviation due only to the vertex corrections in the quark sector defined by $\delta_{q-vertex} = (\Gamma_q-\Gamma_q^{obl})/\Gamma_q^{obl}$. In general, there is a strong dependence of the corrections which can be important in order to evaluate the correct model if a Z' boson is detected. In particular, the vertex corrections introduce negative contributions. This behavior shows that the width decreases if vertex corrections are included. Thus, the vertex corrections reduce the oblique contributions.

The consequences of the corrections of the Z' decay in pp collisions in the CERN LHC collider can be evaluated from figure 5, which displays the ratio of one-loop to tree-level invariant mass distribution for the di-lepton system as a final state, where a central value $M_{Z'} = 1.5$ TeV, and $m_J = 2$ TeV is chosen. As expected, the final state corrections dominate over the resonance range, where the one-loop distribution exhibit bigger values than the tree-level distribution. The biggest deviation is obtained in the peak at $M_{Z'}$,

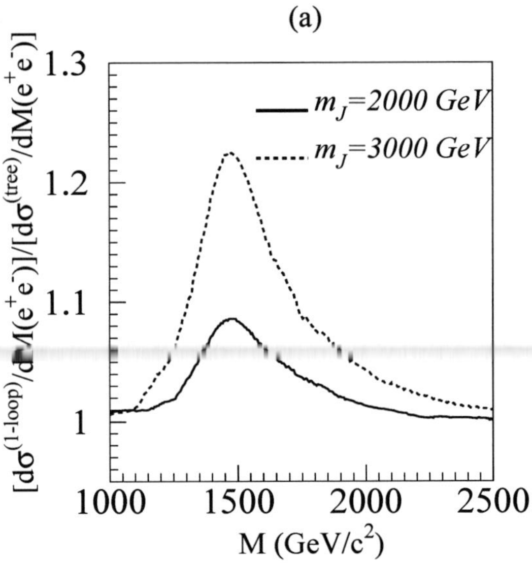

FIGURE 5. Di-electron ratio of the one-loop to tree-level invariant mass distribution at LHC

where the one-loop distribution is 1.09 times the tree-level value for $m_J = 2$ TeV, and 1.23 for $m_J = 3$ TeV [16].

CONCLUSIONS

In the framework of the FLT 331 model, Z' production at LHC was described. For an integrated luminosity of $100 fb^{-1}$ in LHC and considering a central value of $M_{Z'} =$ 1500 GeV, a huge narrow resonance above the SM background is found for di-electron events. Z' signals can be evaluated by studying different final states for fermion and higgs events. One-loop corrections introduce important effects on the Z' decay width, and should be taken into account in order to distinguish this model from other models. For further details see refs. [15, 16]

REFERENCES

1. For reviews, see J. Hewett and T. Rizzo, Phys. Rept. **183**, 193 (1989); A. Leike, Phys. Rept. **317**, 143 (1999); T. Rizzo, hep-ph/0610104; P. Langacker, hep-ph/0801.1345.
2. J. Alcaraz et al. [ALEPH, DELPHI, L3, OPAL Collaborations, LEP Electroweak Working Group], hep-ex/0612034
3. A. Abulencia et al. [CDF Collaboration], Phys. Rev. Lett. **96**, 211801 (2006); ibid **95**, 252001 (2005); ibid **96**, 211802 (2006).
4. T. Aaltonen et al. [CDF Collaboration], Phys. Rev. Lett. **99**, 171802 (2007).
5. M. Dittmar, A.S. Nicollerat and A. Djouadi, Phys. Lett. **B583**, 111 (2004)
6. S. Godfrey, Phys. Rev. **D51**, 1402 (1995); T. G. Rizzo, hep-ph/0303056.
7. R. N. Mohapatra, Unification and Supersymmetry, (Springer, New York, 1986).

8. N. Arkani-Hamed, A. G. Cohen and H. Georgi, Phys. Lett. **B513**, 232 (2001).
9. F. Pisano and V. Pleitez, Phys. Rev. **D46**, 410 (1992); P.H. Frampton, Phys. Rev. Lett. **69**, 2889 (1992); R. Foot, O.F. Hernandez, F. Pisano, V. Pleitez, Phys. Rev. **D47**, 4158 (1993); V. Pleitez and M.D. Tonasse, Phys. Rev. **D48**, 2353 (1993); P.H. Frampton, P. Krastev and J.T. Liu, Mod. Phys. Lett. **9A**, 761 (1994); P.H. Frampton et. al. Mod. Phys. Lett. **9A**, 1975 (1994) Nguyen Tuan Anh, Nguyen Anh Ky, Hoang Ngoc Long, Int. J. Mod. Phys. **A16**, 541 (2001).
10. R. Foot, H.N. Long and T.A. Tran, Phys. Rev. **D50**, R34 (1994); H.N. Long, ibid. **53**, 437 (1996); *ibid*, **54**, 4691 (1996); Mod. Phys. Lett. A **13**, 1865 (1998).
11. J.S. Bell, R. Jackiw, Nuovo Cim. **A60** 47 (1969); S.L. Adler, Phys. Rev. **177**, 2426 (1969); D.J. Gross, R. Jackiw, Phys.Rev. **D6** 477 (1972). H. Georgi and S. L. Glashow, Phys. Rev. **D6** 429, (1972); S. Okubo, Phys. Rev. **D16**, 3528 (1977); J. Banks and H. Georgi, Phys. Rev. **14** 1159 (1976).
12. P.H. Frampton, in Procc. Particles, Strings, and Cosmology (PASCOS) eds. K.C. Wali, Syracuse, NY, 1994, arXiv:hep-ph/9409331.
13. C.A.de S. Pires and O.P. Ravinez, Phys. Rev. **D58**, 35008 (1998); P.V. Dong and H.N. Long, Int. J. Mod. Phys. **A21**, 6677 (2006); C.A. de S. Pires, Phys. Rev. **D60**, 075013 (1999).
14. R. D. Peccei and H. Quinn, Phys. Rev. Lett. **38**, 1440 (1977); Phys. Rev. **D16**, 1791 (1977); P. B. Pal, Phys. Rev. **D52**, 1659 (1995)
15. J.G. Dueñas, N. Gutierrez, R. Martinez, F. Ochoa, Eur. Phys. J. **C60**, 653 (2009).
16. R. Martinez, F. Ochoa, hep-ph/0909.1121 (2009).
17. L. D. Ninh, H. N. Long, Phys. Rev. **D72**, 075004 (2005); E. Ramirez Barreto, Y.A. Coutinho, J. Sa Borges, Phys. Lett. **B632**, 675 (2006); Eur. Phys. J. **C50**, 909 (2007); Nucl. Phys. **B810**, 210 (2009); R. Martinez, F. Ochoa, Eur. Phys. J. **C51**, 701 (2007).
18. Rodolfo A. Diaz, R. Martinez, F. Ochoa, Phys. Rev. **D69**, 095009 (2004).
19. Rodolfo A. Diaz, R. Martinez, F. Ochoa, Phys. Rev. **D72,** 035018 (2005).
20. A. Carcamo, R. Martinez and F. Ochoa, Phys. Rev. **D73,** 035007 (2006).

Radiative Corrections to Asymmetry Parameter in the $\Omega^- \to \Lambda + K^-$ Decay

A. Queijeiro

Departamento de Física, Esc. Sup. de Física y Matemáticas Instituto Politécnico Nacional, U.P. Adolfo López Mateos, México D.F. 07738, email:`aquei@esfm.ipn.mx`

Abstract. We compute the radiative corrections, to first order in the fine structure constant α, to the asymmetry parameter α_Ω of the $\Omega^- \to \Lambda + K^-$ decay. We use previous results where Sirlin's procedure is used to separate the radiative corrections into two parts, one independent model contribution and a model dependent one.

Keywords: Ω^- decay, asymmetry parameter, radiative corrections
PACS: 12.15.Lk,13.40.Ks,13.30.-a,13.30.Eg

A la memoria del maestro y amigo Dr. Augusto García González

Among the decay modes of the Ω-hyperon the most favored is $\Omega^- \to \Lambda^0 + K^-$ with 67.8 %, followed by $\Omega^- \to \Xi^0 + \pi^-$ with 23.6 % and by $\Omega^- \to \Xi^- + \pi^0$ with 8.6 % [1]. All of the particles involved in these decays are hadrons, and the first two of them are quite similar to each other: the final state consists of a neutral baryon and a charged pseudoscalar meson. The decay amplitude for the diagram, shown in Fig. 1, can be written in terms of two form factors as follows:

$$\mathfrak{M}_0 = \overline{u}(q_2) W^\mu(p, q_2) u_\mu(p), \tag{1}$$

where $u(q_2)$ and $u_\mu(p)$ are Dirac spinor of the neutral baryon and the Rarita-Schwinger spin-vector of the Ω hyperon, respectively. $W^\mu(p,q_2)$ is the weak vertex, which takes the general form

$$\begin{aligned} W^\mu(p,q_2) &= a_1\gamma^\mu + a_2 p^\mu + a_3 q_2^\mu \\ &\quad + \left(b_1\gamma^\mu + b_2 p^\mu + b_3 q_2^\mu\right)\gamma_5 \\ &\quad + (c + d\gamma_5) i\varepsilon^{\mu\alpha\beta\lambda}\gamma_\alpha p_\beta q_{2\lambda}. \end{aligned} \tag{2}$$

After substituting (2) in (1) the resulting expression can be simplified, taking into account that the Rarita-Schwinger spin-vector is constrained to satisfy

$$\begin{aligned} \gamma^\mu u_\mu(p) &= 0, \\ p^\mu u_\mu(p) &= 0 \end{aligned}$$

and also using the relationship

$$i\varepsilon^{\mu\alpha\beta\lambda}\gamma_\alpha = (g^{\beta\lambda}\gamma^\mu - g^{\beta\mu}\gamma^\lambda + g^{\lambda\mu}\gamma^\beta - \gamma^\beta\gamma^\lambda\gamma^\mu)\gamma_5,$$

CP1259, *First CINVESTAV-UNAM Symposium on High Energy Physics*
edited by J. C. D'Olivo, R. López-Fernández, and M. Á. Pérez

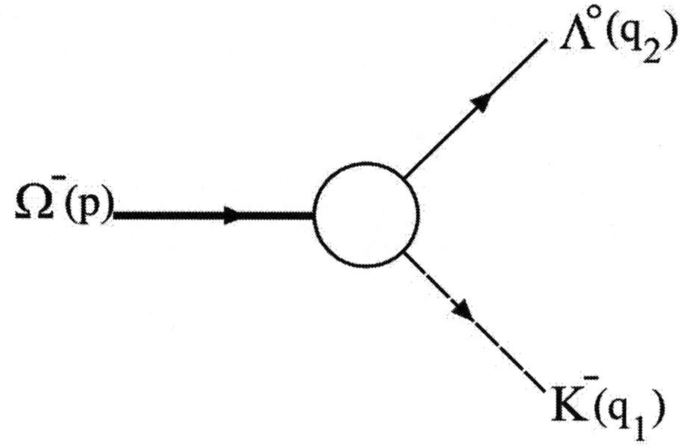

Fig. 1

to

$$\begin{aligned}\mathfrak{M}_0 &= \overline{u}(q_2)\left[(a_3+b_3\gamma_5)q_2^{\mu}\right.\\ &\quad \left.+M_{\Omega}(c+d\gamma_5)\right]q_2^{\mu}u_{\mu}(p)\\ &\equiv \frac{i}{M_{\Omega}}\overline{u}(q_2)(A+B\gamma_5)q_2^{\mu}u_{\mu}(p).\end{aligned} \tag{3}$$

The form factors A and B are the P- and D- wave amplitudes or form factors, related to the values 1 and 2 of the angular momentum final state. These form factors are dimensionless since we have divided them by the only dimensional parameter M_{Ω}, the mass of the Ω particle. The width is

$$\begin{aligned}\Gamma_0 &= \frac{1}{2\pi}\frac{M_B}{M_{\Omega}^2}\widehat{q}_2\sum_{\text{spin}}|\mathfrak{M}_0|^2\\ &= \frac{1}{46\pi}\frac{q_2^3}{M_{\Omega}^2}\left\{|A|^2\left[(M_{\Omega}+M_B)^2-M_P^2\right]\right.\\ &\quad \left.+|B|^2\left[(M_{\Omega}-M_B)^2-M_P^2\right]\right\},\end{aligned} \tag{4}$$

where $M_{B(P)}$ is the mass of the baryon (meson), and q_2 is the magnitude of the baryon 3-momentum, which is equal in magnitude to the meson 3-momentum q_1. (To obtain (4) we use Eq. (6) below, for the sum over spins). Factors in square brackets have values 7.52×10^6 MeV^2 and 6.43×10^4 MeV^2 for the mode $\Omega^-\to\Lambda^0+K^-$, and 8.90×10^6 MeV^2 and 1.08×10^5 MeV^2 for the mode $\Omega^-\to\Xi^0+\pi^-$. Thus, the form factor B is suppressed by one or two orders of magnitude as compared to the form factor

A, except if B itself is bigger than A. However, this turns out not to be the case since, the other observable in the decay $\Omega^- \to \Lambda + K^-$, the asymmetry parameter α_Ω, defined by

$$\alpha_\Omega = \frac{2\mathrm{Re}(A^* B)}{|A|^2 + |B|^2}, \tag{5}$$

has been recently measured by the HyperCP Collaboration [2] for the $\Omega^- \to \Lambda^0 + K^-$ decay, and the reported result shows that α_Ω is small, but non zero and predominantly parity conserving, i.e., dominated by the P-wave final state.

In addition to the weak interaction mediating this decays, there are final-state strong interactions between the outgoing baryon and pion. These final-state interactions are important when comparing decay rates, and the α parameters, among neutral and charged modes in nonleptonic baryon decays. There, strong phase-shifts are taken into account in the S and P amplitudes [3]. For the $\Omega^- \to \Lambda^0 + K^-$ decay, the final state has isospin 1/2, which means we need one strong phase shift only, to describe the P and D amplitudes, having no effect on α_Ω. (For $\Omega^- \to \Xi^0 + \pi^-$ the final state have isospin values 1/2 and 3/2, requiring two phase shifts to describe the P and D amplitudes). In [4] a chiral Lagrangian approach is used to calculate these phases, concluding that the D phase is much smaller than the P one.

In the present work, we are interested in the radiative corrections to the asymmetry parameter α_Ω in $\Omega^- \to \Lambda^0 + K^-$. Since this observable depends only on $|A|^2$ and $|B|^2$ no effect on the phase-shifts is considered. We believe that, in order to provide reliable experimental values of this parameter, knowledge of the radiative corrections is required. We follow Sirlin's procedure introduced in a seminal paper [5], in which he showed that the virtual radiative corrections to neutron decay can be separated into two parts: a model independent one, finite in the ultraviolet region and containing the infrared divergence, and a model dependent one containing all the complications due to strong interactions and the intermediate vector boson and the ultraviolet divergence. The separation procedure is gauge invariant. Later on, the method was generalized, by García and collaborators [6], to other observables in neutron decay and to other hadron semileptonic decays.

Radiative corrections to nonleptonic baryon decays were first performed more than four decades ago by some authors. For example, in their study of the kaon decays $K_S^0 \to \pi^+\pi^-, \pi^0\pi^0$ Abbud, Lee and Yang [3] gave an estimation of the electromagnetic corrections to the real part of A_2/A_0, where A_2 and A_0 are the amplitudes for the charged and neutral modes, respectively. Also, Belavin and Narodetsky [7] calculated the radiative corrections to $\Lambda^0 \to N\pi$ and to $K_S^0 \to 2\pi$, and Jarlskog [8] calculated the radiative corrections to the non-leptonic Σ decays. All of these calculations were done previously to Sirlin's method. Thus, they contains a logarithmically divergent term in the ultraviolet cutoff, and a finite term evaluated exactly if the strong interaction is switched off.

In what follows we present the results for the virtual and bremsstrahlung radiative corrections to the $\Omega^- \to \Lambda^0 + K^-$ decay, previously reported in [9] and use them to calculate the radiative corrections to α_Ω.

We assume that the polarisation of the Ω is $\hat{s}$, then from (3) we obtain for the uncorrected transition amplitude where $\overline{|\mathfrak{M}_0|^2}$ is the squared amplitude summed over polarisations of all particles. We have used the following set of Feynman rules:

the spin-energy projector operator of a spin-3/2 particle [10],

$$\begin{aligned}
\sum(p)^{\mu\nu} &= u^\mu(p)u^\nu(p) \\
&= \frac{m+\gamma^\alpha p_\alpha}{24m}\Big\{-3g^{\mu\nu}+\gamma^\mu\gamma^\nu+\frac{2p^\mu p^\nu}{m^2}+\frac{p^\mu\gamma^\nu-p^\nu\gamma^\mu}{m} \\
&+3\left[-\frac{i}{m}\varepsilon^{\mu\nu\lambda\tau}p_\lambda\eta_\tau+\frac{4}{5}I^{\mu\nu}\gamma_5\gamma^\tau\eta_\tau-\frac{1}{5}\gamma_5\gamma_\rho(I^{\rho\mu}\eta^\nu+I^{\rho\nu}\eta^\mu)\right] \\
&-\left[\eta^{\mu\nu}-\gamma_\rho(\eta^{\rho\mu}\gamma^\nu-\eta^{\rho\nu}\gamma^\mu)-\frac{1}{m}\gamma_\rho(\eta^{\rho\mu}p^\nu-\eta^{\rho\nu}p^\mu)\right] \\
&-\frac{2}{3}\gamma_5\gamma_\rho\eta^{\mu\nu\rho}\Big\},
\end{aligned} \tag{6}$$

where

$$\eta^{\mu\nu} = 3\eta^\mu\eta^\nu - I^{\mu\nu}, \tag{7a}$$

$$\eta^{\mu\nu\rho} = \frac{9}{2}\eta^\mu\eta^\nu\eta^\rho-\frac{9}{10}(I^{\rho\mu}\eta^\nu+I^{\rho\nu}\eta^\mu+I^{\mu\nu}\eta^\rho), \tag{7b}$$

$$I^{\mu\nu} = -g^{\mu\nu}+\frac{p^\mu p^\nu}{m^2}, \tag{7c}$$

η^μ is the spin-vector. The spin-1/2 the spin-energy projector is given by [10]

$$u(p)\overline{u}(p)=\frac{m+\gamma^\alpha p_\alpha}{4m}\left[1+\gamma_5\gamma_\mu\eta^\mu\right]. \tag{8}$$

For the computation of the radiative corrections we use for the Ω - hyperon propagator [11]:

$$\begin{aligned}
\Sigma_{\mu\nu}(p) &= \Big\{\Lambda_{\mu\alpha}\Big[g^{\alpha\beta}(i\partial\!\!\!/-M_\Omega)+\frac{i}{3}(\gamma^\alpha\partial\!\!\!/\gamma^\beta-\gamma^\alpha\partial^\beta-\gamma^\beta\partial^\alpha) \\
&+\frac{1}{3}M_\Omega\gamma^\alpha\gamma^\beta\Big]\Lambda_{\beta\nu}\Big\}^{-1} \\
&= \Delta_{\beta\nu}D^{\alpha\beta}(p)\Delta_{\mu\alpha}
\end{aligned} \tag{9}$$

with

$$\begin{aligned}
\Delta_{\beta\nu} &= g_{\beta\nu}-\frac{1}{3}\gamma_\beta\gamma_\nu, \\
D^{\alpha\beta}(p) &= \frac{1}{p^2-M_\Omega^2}\Big\{\Big[-g^{\alpha\beta}+\frac{1}{3}\gamma^\alpha\gamma^\beta-\frac{1}{3M_\Omega}(\gamma^\alpha p^\beta-\gamma^\beta p^\alpha) \\
&+\frac{2}{3M_\Omega^2}p^\alpha p^\beta\Big](p\!\!\!/+M_\Omega) \\
&-\frac{2}{3}\frac{p^2-M_\Omega^2}{M_\Omega^2}(\gamma^\alpha p^\beta+\gamma^\beta p^\alpha-\gamma^\alpha p\!\!\!/\gamma^\beta+M\gamma^\alpha\gamma^\beta\Big\}
\end{aligned} \tag{10}$$

and

$$\Lambda_{\mu\alpha}(D) = g_{\mu\alpha} - \gamma_\mu \gamma_\alpha \tag{11}$$

is the contact transformation, with parameter $D = -1$. Under the minimal coupling rule $\partial_\mu \to \partial_\mu - ieA_\mu$, the $\Omega\gamma\Omega$ electromagnetic vertex is

$$ie\Gamma_{\nu\mu\alpha} = \Lambda_{\nu\lambda}\Gamma^{\lambda\sigma}_{0\mu}\Lambda_{\sigma\alpha}, \tag{12}$$

where

$$\Gamma^{\lambda\sigma}_{0\mu} = g^{\lambda\sigma}\gamma_\mu + \frac{1}{3}\gamma^\lambda\gamma_\mu\gamma^\sigma - \frac{1}{3}(\gamma^\lambda g^\sigma_\mu + \gamma^\sigma g^\lambda_\mu). \tag{13}$$

The index μ in (12) is associated to the photon field A_μ, while the subindex 0 is included since this vertex will be modified later, because this vertex rule does not give the correct value for the magnetic moment. We can add to (12) some terms linear in the photon 4-vector momentum k

$$\Gamma^{\lambda\sigma}_{\mu} = \Gamma^{\lambda\sigma}_{0\mu} + \frac{\kappa_\Omega}{M_\Omega}(k^\lambda g^\sigma_\mu - k^\sigma g^\lambda_\mu) + \frac{\lambda_\Omega}{M_\Omega} g^{\lambda\sigma}(\not{k}\gamma_\mu - k_\mu). \tag{14}$$

Nevertheless, only the first summand in (13) and the first summand in (10) is considered to compute the independent model radiative corrections, as is explained in [9].

For the $\Omega(p_\beta) - \Lambda(q_2) - K(q_1)$ vertex we have

$$W_\beta(q_1) = \frac{i}{M_\Omega}(A + B\gamma_5)\Lambda_{\beta\nu}q_1^\nu \tag{15}$$

and for the $\Omega(p_\beta) - \Lambda(q_2) - K(q_1) - \gamma(k_\mu)$ vertex

$$V_{\beta\mu} = -\frac{ie}{M_\Omega}(A + B\gamma_5)\Lambda_{\beta\mu}. \tag{16}$$

Consider first the virtual radiative corrections. The independent model contribution comes from the first three diagrams of the set of diagrams in Fig. 2. As is shown in [9], following Sirlin's method the amplitude of this set of diagrams is

$$\mathfrak{M}_{(V)i} = \frac{\alpha}{\pi}\phi\mathfrak{M}_0 + \frac{\alpha}{\pi}\frac{E_1}{M_1^2}\phi'\bar{u}(q_2)W^\beta(q_1)\,\not{q}_1 u_\beta(p). \tag{17}$$

where

$$\phi = -\frac{1}{2}\ln\left(\frac{\lambda^2}{M_\Omega^2}\right)(1+\frac{1}{2\beta_1}\ln\frac{1-\beta_1}{1+\beta_1})$$
$$+\frac{1}{8\beta_1}\ln^2\frac{1-\beta_1}{1+\beta_1}-\frac{1}{2}(\frac{3}{2}-\frac{M_\Omega^2}{M_2^2})\ln\frac{M_1}{M_\Omega}$$
$$+\left[\frac{-M_1^2+M_\Omega E_1(1+\beta_1^2)}{2M_2^2}\right]\frac{1}{2\beta_1}\ln\frac{1-\beta_1}{1+\beta_1}$$
$$-\frac{1}{2\beta_1}\left[Li_2(\frac{b_1-b_2}{b_2})-Li_2(-\frac{b_1-b_2}{b_2-1})\right]$$
$$-\frac{1}{\beta_1}\ln\left(\frac{b_1-1}{b_2-1}\right)\left[-\ln\frac{M_1}{M_\Omega}+\frac{1}{2}\ln\frac{1-\beta_1}{1+\beta_1}\right.$$
$$\left.+\frac{1}{2}\ln\left(\frac{b_1-1}{b_2-1}\right)\right]-\frac{11}{16}, \tag{18}$$

$$\phi' = \frac{1-\beta_1^2}{\beta_1}\left[-\frac{M_2^2+M_\Omega E_1-M_\Omega^2}{2M_2^2}\tanh^{-1}\beta_1-\frac{M_\Omega q_1}{2M_2^2}\ln\frac{M_\Omega}{M_1}\right], \tag{19}$$

$$b_{1,2} = \frac{M_\Omega}{M_2^2}\left[M_\Omega - E_1(1\mp\beta_1)\right] \tag{20}$$

with $\beta_1 = q_1/E_1$ the velocity of the meson particle, in units of the speed of light, $Li_2(z)$ is Spence function and λ is the fictitious photon mass. From (17) we compute the contribution to the transition rate through its combination with the uncorrected amplitude:

$$|\mathfrak{M}_0|^2+2\mathrm{Re}(\overline{\mathfrak{M}}_0\mathfrak{M}_{(V)i}) = \overline{|\mathfrak{M}_0|^2}\left\{1+\alpha_\Omega^V\hat{s}\cdot\hat{q}_1+\ldots\right\}, \tag{21}$$

where

$$\alpha_\Omega^V = \left[1+\frac{6\alpha}{q_1^2}(\phi+\frac{M_\Omega E_1}{M_K^2}\phi')\right]\alpha_\Omega^0. \tag{22}$$

We have aligned the spin of the decaying hyperon such that $\hat{s}\cdot\hat{q}_1 = \cos\theta$. Thus, the contribution to the angular transition rate is

$$\frac{d\Gamma_V}{d\cos\theta} = \Gamma_0'(1-\alpha_\Omega^V\cos\theta+\ldots). \tag{23}$$

Here Γ_0' is the modified transition rate after taking into account the virtual dependent model contribution in the following way. We cannot evaluate explicitly the model dependent part, but the general form it takes is

$$\mathfrak{M}_{(V.MD)} = \frac{i\alpha}{M_\Omega}\overline{u}(q_2)(F+G\gamma_5)q_1^\beta u_\beta(p), \tag{24}$$

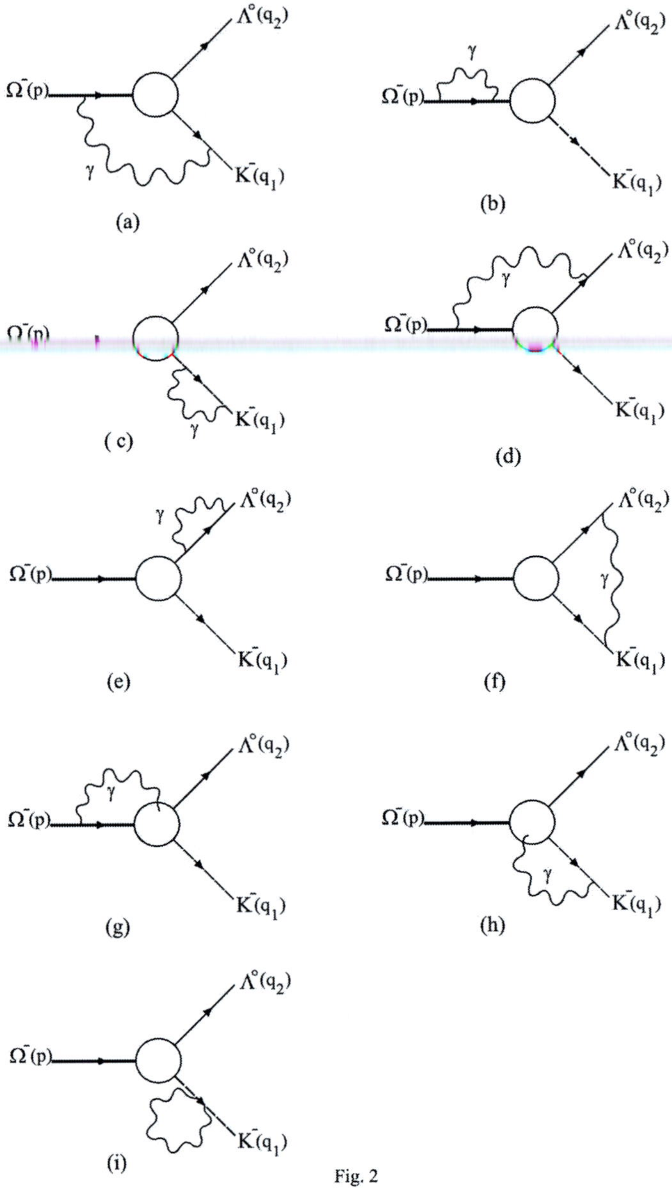

Fig. 2

which can be added to the uncorrected amplitude in (3) and define new form factors

$$A' = A + \frac{\alpha}{\pi}F,$$
$$B' = B + \frac{\alpha}{\pi}G.$$

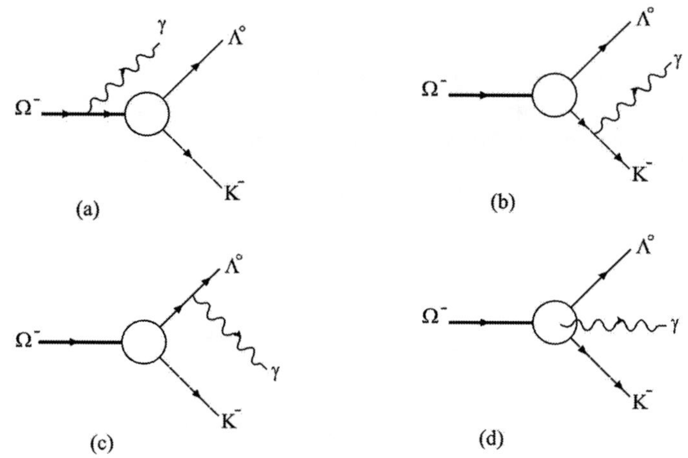

Fig. 3

Then, neglecting terms of second order in α,

$$\Gamma_0' \approxeq \Gamma_0 + \Gamma_V^d = \frac{q_1^3}{24\pi M_\Omega^4}\left\{\left[(M_\Omega + M_2)^2 - M_1^2\right]|A'|^2 + \left[(M_\Omega - M_2)^2 - M_1^2\right]|B'|^2\right\}. \tag{25}$$

Now we proceed to consider the bremsstrahlung corrections. In Fig. 3 we show the set of diagrams contributing to the bremsstrahlung contribution. The model independent contribution is given by

$$|\mathfrak{M}_B|^2 = 4\pi\alpha\beta_1^2\overline{|\mathfrak{M}_0|^2}\left(1 + \alpha_\Omega^0\frac{\hat{s}\cdot\vec{q}_1}{|\vec{q}_2|} - \frac{2\alpha_\Omega^0}{5}\frac{\hat{s}\cdot\vec{k}}{|\vec{q}_2|}\right) \times \frac{1-(k^2/\omega^2)(\hat{k}\cdot\hat{q}_1)^2}{(\omega - k\beta_1\hat{k}\cdot\hat{q}_1)^2}. \tag{26}$$

Here ω is photon energy and $\hat{k}$ is the direction of its 3-momentum vector. We can extract the infrared divergent part

$$\frac{d\Gamma_B}{d\cos\theta} = \frac{\alpha}{4\pi}\Gamma_0(1 + \alpha_\Omega \cos\theta)\Theta, \tag{27}$$

where

$$\Theta = \ln\left(\frac{2k_m}{\lambda}\right)\left(-2 + \frac{2}{\hat{\beta}_1}\tanh^{-1}\beta_1\right) + 1 + \frac{1}{2\beta_1}\tanh^{-1}\beta_1\left[2 + \ln\left(\frac{1-\beta_1^2}{4}\right)\right] + \frac{1}{\beta_1}\left[Li_2(\beta_1) - Li_2(-\beta_1)\right] + \frac{1}{2\beta_1}\left[Li_2\left(\frac{1-\beta_1}{2}\right) - Li_2\left(\frac{1+\beta_1}{2}\right)\right]. \tag{28}$$

Defining

$$\alpha_\Omega^B = \frac{\alpha}{2\pi}\Gamma_0\alpha_\Omega\,\Theta \tag{29}$$

we can write

$$\frac{d\Gamma_B}{d\cos\theta} = \frac{1}{2}\Gamma_0(\frac{\alpha}{2\pi} + \alpha_\Omega^B \cos\theta). \tag{30}$$

The radiative corrections to α_Ω, when hard photons with energies above k_m are experimentally discriminated, is the sum of (23) and (30)

$$\frac{d\Gamma}{d\cos\theta} = \frac{1}{2}\Gamma_0' \left[1 + \frac{\alpha}{2\pi}(2\phi + \Theta) + \widetilde{\alpha}_\Omega \cos\theta\right], \tag{31}$$

where

$$\widetilde{\alpha}_\Omega = \alpha_\Omega^V + \alpha_\Omega^B \tag{32}$$

Is the asymmetry parameter with radiative corrections up to order α.

ACKNOWLEDGMENTS

The author wishes to acknowledge partial support from COFAA- and SIP- IPN, and also to Dr. A. Martínez for stimulating conversations.

REFERENCES

1. C. Amsler, et al., *Phys. Lett.* **B667**, 1 (2008).
2. W.-M. Yao et al., *Journal of Physics G* **33**, 1 (2006); W. Chan et el., *Phys. Rev D* **58**, 072002 (1998).
3. F. Abbud, B. Lee, and C. Yang, *Phys. Rev. Lett.* **18**, 980 (1967).
4. See for instance C.C. Barros Jr., hep-ph/0303237.
5. A. Sirlin, *Phys. Rev.* **164**, 1767 (1967).
6. See, for instance, A. García and M. Maya, *Phys. Rev. D* **23**, 2603 (1981) for $M_{l3}^{\pm}$ decays, A. García and M. Maya, *Phys. Rev. D* **17**, 1376 (1978) for polarizad-neutron decay, A. García and S.R. Juárez, *Phys. Rev. D* **22**, 1132 (1980) for semileptonic decays of unpolarizdd hyperons, and S.R. Juárez, A. Martínez and A. García *Phys. Rev. D* **35**, 232 (1987) for polarized hyperons.
7. A. Belavin, and L. Narodetsky, *Phys. Lett.* **B26**, 668 (1968).
8. C. Jarlskog, *Nucl. Phys.* **B3**, 365 (1967).
9. A. Queijeiro (2010), submitted for publication in Physica Scripta.
10. S.Y. Choi, T. Lee, and H.S. Song, *Phys. Rev. D* **D40**, 2477 (1989).
11. T. Pilling, *Int. J. Mod. Phys.* **A20**, 2715 (2005).

THEORY

Quantal Time Asymmetry: Mathematical Foundation and Physical Interpretation

A. Bohm

Department of Physics, University of Texas at Austin, Austin, TX 78712, U.S.A.

Abstract. Time in standard quantum mechanics extends from $-\infty < t < +\infty$; this is the result of a mathematical theorem (Stone-von Neumann) for the solutions of the Schrödinger equation for states or of the Heisenberg equation for observables. In reality t does not extend to $t \to -\infty$, since according to causality, a quantum state ϕ^+ must be prepared first at a particular time $t = t_0$, before the probability $|(\psi^-(t), \phi^+(t_0))|^2$ for an observable ψ^- can be measured in it at $t > t_0$ (Feynman (1948)). In experiments on single Ba^+ ions, Dehmelt and others observed this finite preparation time as the ensemble of onset-times $t_0^1, t_0^2, \ldots, t_0^n$ of dark periods. How the semigroup time evolution, $t_0 \equiv 0 < t < \infty$ with a beginning of time t_0, can suggest the parametrization of the resonance pole position of the Z-boson at $s = s_R$ as $s_R = (M_R - i\Gamma_R/2)^2$ in terms of a mass M_R and a width Γ_R given by a lifetime $\tau = \hbar/\Gamma_R$, is the subject of this contribution dedicated to Augusto García.

Keywords: rigged Hilbert space, resonances, Z boson, time asymmetry
PACS: 03.65.Db, 03.65.Nk, 11.10.St, 14.70.Hp

INTRODUCTION

The definition of mass and width of a relativistic resonance has been a problem for a long time. For some time the Particle Data Group [1] listed two or three different values for the mass of the Z-boson and some of the hadron resonances obtained from the same experimental data with different (theoretical) definitions for the lineshape of a resonance. It was Augusto García who told me, during one of my visits at CINVESTAV[1], that the standard definition (M_Z, Γ_Z) for the Z-boson, based on the on-mass-shell renormalization scheme, is not gauge invariant (NNLO, $\mathcal{O}(g^6)$ and $\mathcal{O}(g^4)$) [2]. This started me on the problem of the *definition of relativistic resonances*.

There are two different definitions of the Z-boson lineshape

For the propagator definition in the on-mass-shell-renormalization scheme:

$$a_j^{\text{res}} = \frac{R_Z}{s - M_Z^2 + i\frac{s}{M_Z}\Gamma_Z} \qquad \text{(1a)}$$

For the S-matrix pole definition:

$$a_j^{\text{res}} = \frac{r}{s - s_R} \qquad \text{(1b)}$$

---

[1] Centro de Investigación y de Estudios Avanzados del Instituto Politécnico Nacional, Mexico

CP1259, *First CINVESTAV-UNAM Symposium on High Energy Physics*
edited by J. C. D'Olivo, R. López-Fernández, and M. Á. Pérez

Here s_R is the pole position of the complex S-matrix pole, and

$$a_j^{\text{res}} = \frac{r}{s - \overline{M}_Z^2 + i\overline{M}_Z\overline{\Gamma}_Z} \tag{2}$$

$$a_j^{\text{res}} = \frac{r}{s - (M_R - i\Gamma_R/2)^2} \tag{3}$$

are two different parametrizations of the pole position s_R.

Different "experimental" values of (M,Γ) are obtained from fits of the partial wave amplitude of angular momentum j,

$$a_j(s) = a_j^{\text{res}}(s) + B_j(s) \tag{4}$$

to the cross section data ("line shape") $|a_j(s)|^2$ (and also to the asymmetries) [2].

The different parametrizations of the resonance amplitude (1), (2), (3) lead to different "experimental" values of the resonance mass and width[3]:

$$\begin{aligned} M_Z &= 91.1875 \pm 0.0021\,\text{GeV} & \Gamma_Z &= 2.4939 \pm 0.0024\,\text{GeV} \\ M_R &= 91.1611 \pm 0.0023\,\text{GeV} & \Gamma_R &= 2.4943 \pm 0.0024\,\text{GeV} \\ \overline{M}_Z &= 91.1526 \pm 0.0023\,\text{GeV} & \overline{\Gamma}_Z &= 2.4945 \pm 0.0024\,\text{GeV} \\ M_Z - M_R &= 0.026 \pm 0.004\,\text{GeV} & M_Z - \overline{M}_Z &= 0.035 \pm 0.004\,\text{GeV} \end{aligned} \tag{5}$$

This differences between the different mass values are significant as compared to the experimental errors $\sigma_{M_Z} = 0.0021\,\text{GeV}$, and therefore one may ask the question: Which of these (M,Γ) is the right mass and width?

The same situation also holds for the different mass definitions of the hadron resonances for which sufficiently accurate data are available, discussed extensively at CINVESTAV [4]

$$M_\Delta - \overline{M}_\Delta \approx 19\,\text{MeV} \quad (\sim 2\% \text{ of } M_\Delta) \qquad M_\rho - \overline{M}_\rho \approx 10\,\text{MeV} \quad (\sim 1.5\% \text{ of } M_\rho)$$

Thus one has again the same important question: What is the right definition of mass M and width Γ?

This leads to more general questions: What is the a resonance? Does a resonance have a lifetime even if one cannot measure it? In non-relativistic (atomic, molecular) physics, the majority opinion is: a resonance of width Γ and a decaying state of lifetime τ are just different manifestations of the same state and $\tau = \hbar/\Gamma$. The opposite opinion dominates in relativistic particle physics: a lifetime τ for broad relativistic resonances is not defined.

In these lecture notes we want to address a more positive question: can the relation $\tau = \hbar/\Gamma$ between an experimental lifetime τ of a decaying state and a width Γ of

---

[2] The amplitude $B_j(s)$ is slowly varying near the singularity of (1); $B_j(s)$ describes the non-resonant background (or contribution from other, far away, resonances.

[3] M_Z, $\overline{M}_Z$ and M_R were listed in PDG book [1], now only (M_Z, Γ_Z) [3] is listed as the standard value.

the resonance help us to find the right definition for mass and width of a relativistic resonance? Which of the parametrizations used in (1), (2), (3) or any other thinkable parametrizations fulfills the relation $\tau = \hbar/\Gamma$? The answer to this question would unify the theory of relativistic resonances and decaying states.

The starting point for a relativistic theory of particles and fields [5] is Wigner's definition of stable relativistic particles by unitary irreducible representations (m, j) of the Poincaré group (the invariance group of relativistic space-time). In analogy to this definition of stable relativistic particles by the unitary representations (m =real mass, j =integer, or half integer spin), a quasistable relativistic particle should be defined by a *semigroup* representation of the Poincaré group. The semigroup representation is characterized by $(s_R \equiv M_R - i\Gamma_R/2, j)$; here j is again spin and s_R is the position of the resonance pole of the j–th partial S-matrix element $S_j^{\eta'\eta}(s)$, where $s = p_\mu p^\mu$. The invariant s can be analytically continued into the complex plane second sheet, and $\eta'\eta$ are the particle species numbers [5] or channel quantum numbers of the in- and out-configuration. The two parameters (M, Γ) or $(M, \hbar/\tau)$ are also the two parameters by which one catalogs the relativistic resonances or the decaying states in [1]. However in the relativistic case it is not quite clear how one should divide the j–th partial wave amplitude (4) into an amplitude specific to the resonance $a_j^{\rm res}(s)$ and the rest $B_j(s)$. This is the problem discussed in these lecture notes.

RESONANCES AND DECAYING STATES

For the sake of simplicity, and since the talk was addressed to a broader audience we want to discuss here mainly the non-relativistic case. We show how a physically minor but mathematically significant modification of quantum theory leads to $\tau = \hbar/\Gamma$ as an exact equality for the non-relativistic case, and then also for the relativistic case.

Comparing

Resonances

Resonances characterized by (E_R, Γ) appear as Breit-Wigner lineshape in the cross section

$$\sigma_j(E) \sim |a_j(E)|^2 = \left| \frac{r_\eta}{E - (E_R - i\frac{\Gamma}{2})} + B(E) \right|^2 \qquad \text{(6R)}$$

$B(E)$ is a slowly varying function of E (background).

and Decaying States

Decaying states characterized by (E_D, τ), are measured by the exponential law for the counting rate of the decay products η

$$\frac{\Delta N(t_i)}{\Delta t_i} = R_\eta(t) \propto \mathrm{e}^{-\frac{t}{\tau}}, \qquad \text{(6D)}$$

here $\Delta N(t_i)$ is the number of decay products registered in the detector during the time interval Δt_i around t_i. t_i is the lifetime of each *individual* decaying particle created at $t = 0 (= t_0)$

The fit of the Z-boson resonance data with $\sqrt{s} = E$ exhibits the Breit-Wigner

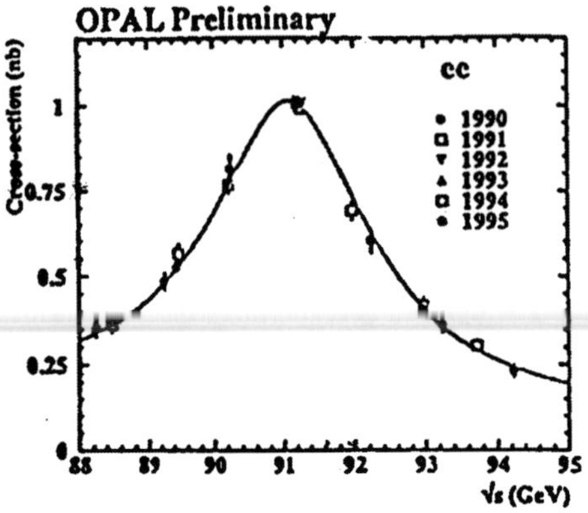

FIGURE 1.

For the K_S^0 decay rate (with the K_L^0 interference analyzed out [6]) the time dependence is exponential

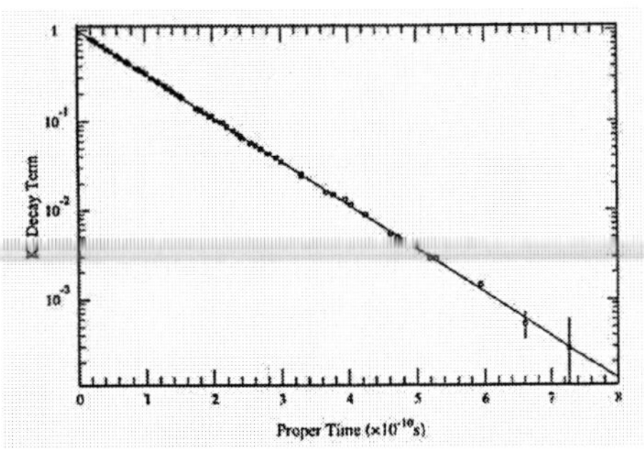

FIGURE 2.

Many people think that resonances are the same as decaying states, and especially for non-relativistic quantum mechanics, a common assumption is that

$$\frac{\hbar}{\Gamma} = \tau, \quad \left(\text{or at least } \frac{\hbar}{\Gamma} \approx \tau\right). \tag{7}$$

This relation is based on the Weisskopf-Wigner (W-W) approximation [7].

But in standard quantum theory there is no proof of $\hbar/\Gamma = \tau$. Using W-W methods, the probability to observe the decay products $|\psi\rangle\langle\psi|$ in a prepared resonance state $\phi(t)$ with Breit-Wigner width Γ is derived as [8]

$$|\langle\psi|\phi(t)\rangle|^2 = \mathscr{P}_{\phi(t)}(\psi) \sim \mathrm{e}^{-\Gamma t/\hbar} + \Gamma \times (\text{additional terms}). \tag{8}$$

One can prove that there is no Hilbert space vector $\phi(t)$ that obeys the exact exponential law, i.e., for which (additional terms of (8))=0.

In Fig. 2 for the K_S^0-decay experiment (with K_L^0 interference analyzed out [6]) the time dependence of the decay rate looks perfectly exponential. This means the calculated Born probability for detecting $\pi\pi$ in K_S^0 decay

$$|\langle\psi_{\pi\pi} \mid \phi^{K_S^0}(t)\rangle|^2 = \frac{N(t)}{N}, \tag{9}$$

needs to be a perfect exponential, as shown for $N(t)$ by the straight line in Fig. 2.

The easiest way to derive an exactly exponential decay probability is to *postulate* a state vector ϕ^G which has the property:

$$H\phi^G = (E_R - i\frac{\Gamma}{2})\phi^G \quad \text{(10a)} \qquad \text{and} \qquad \phi^G(t) = \mathrm{e}^{-iHt/\hbar}\phi^G, \quad \text{(10b)}$$

The decay probability into any observable $|\psi\rangle\langle\psi|$ is then obtained by using (10b) and then (10a) in the exponent:

$$\mathscr{P}_{\phi^G(t)}(\psi) = |\langle\psi \mid \phi^G(t)\rangle|^2 = |\langle\psi \mid \mathrm{e}^{-iHt/\hbar} \mid \phi^G(t)\rangle|^2$$
$$= |\langle\psi \mid \phi^G\rangle \mathrm{e}^{-i(E_R - i\Gamma/2)t/\hbar}|^2 = |\langle\psi \mid \phi^G\rangle|^2 \mathrm{e}^{-\Gamma t/\hbar}. \quad (11)$$

Though this leads to the exponential law (11), this vector ϕ^G (called Gamow vector [9]) makes no sense in standard (Hilbert space) quantum mechanics. The reason is that if one solves the Dynamical Equations:

either in the Heisenberg picture for observables $\Lambda(t)$ or $|\psi(t)\rangle\langle\psi(t)|$

$$i\hbar\frac{\partial\Lambda(t)}{\partial t} = -[H, \Lambda(t)] \quad (12a)$$

$$i\hbar\frac{\partial}{\partial t}\psi(t) = -H\psi(t) \quad (12b)$$

or in the Schrödinger picture for states $\phi(t)$

$$i\hbar\frac{\partial}{\partial t}\phi(t) = H\phi(t) \quad (12c)$$

under standard quantum mechanical *boundary conditions*:

$$\text{set of states } \{\phi\} = \mathscr{H} = \text{Hilbert space} = \text{set of observables } \{\psi\} \quad (13a)$$

or

$$\phi(E) = \langle E|\phi\rangle \in L^2(E) \quad \psi(E) = \langle E|\psi\rangle \in L^2(E), \quad E_0 < E < \infty, \quad (13b)$$

then one obtains (by the Stone-von Neumann Theorem) the unitary group evolution:

$$\psi(t) = \mathrm{e}^{iHt/\hbar}\psi, \quad -\infty < t < +\infty \quad \text{(in the Heisenberg picture)} \quad \text{or} \quad \phi(t) = \mathrm{e}^{-iHt/\hbar}\phi, \quad -\infty < t < +\infty \quad \text{(in the Schrödinger picture)}. \quad (14)$$

This would lead to the "exponential catastrophe" for a Gamow vector

$$\mathscr{P}_{\phi^G(t)}(\psi) \sim \mathrm{e}^{-\Gamma t/\hbar} \to \infty \quad \text{if one extends } t \text{ to } -\infty, \quad (15)$$

as dictated by the Stone-von Neumann theorem in (14).

To give a mathematical meaning to a Gamow vector, one needs another theory than the one based on axiom (13). This theory must:

1. extend the energy to discrete and continuous values in the complex energy plane
2. restrict the time to $0 \equiv t_0 \leq t < +\infty$ because causality requires that:
   (a) a K_S^0 must be prepared first at a time $t = t_0$ before one can detect its decay products $\pi^+\pi^-$ at $t > t_0$.
   (b) it avoid the exponential catastrophe (15).

We have become so used to time t extending over $-\infty < t < +\infty$ in (14),

$$U(t) = \mathrm{e}^{iHt/\hbar}; \quad -\infty < t < +\infty, \quad (16)$$

that we forget that the unitary group (16) is a consequence of a mathematical theorem, the Stone-von Neumann theorem for the Hilbert space (i.e., Lebesgue square integrable energy wave function $\phi(E)$). The Hilbert space of Lebesgue square integrable functions [10] was a great achievement for the development of a mathematical theory for quantum mechanics. But as documented by the recent textbooks of quantum mechanics, physicists do not use Lebesgue integration. They simply follow the Dirac formulation which is mathematically justified by the Schwartz space. In addition physicists continued the energy of the Dirac kets into the complex energy plane using Lippmann-Schwinger kets. But they hesitated to accept the Gamow kets like (10a) because $-\infty < t < +\infty$ in (14) and (16) leads to the "catastrophe" (15).

The restriction $t_0 < t < +\infty$ is more natural than $-\infty < t < +\infty$. The finite t_0 can be given the meaning of the time at which the state $\phi^{K_S^0}(t_0)$ in (9) is prepared. And a state $\phi^{K_S^0}(t_0)$ needs to be prepared first before the observable $|\psi_{\pi\pi}(t)\rangle\langle\psi_{\pi\pi}(t)|$ can be detected in this state for $t > t_0$. Physically $N(t)/N$ of (9) can be counted only for times $t_i > t_0$, and thus its theoretical representative $|\langle\phi^{K_S^0}(t_0) \mid \psi_{\pi\pi}(t)\rangle|^2$, needs to exist for $t > t_0$ only.

QUANTUM MECHANICAL TIME ASYMMETRY

Quantum Mechanical "Beginning of Time"

From the remarks at the end of the previous section the experimental situation is much better described by a time evolution *semi-group* $\mathscr{U}(t)$ with $t_0 = 0 \leq t < \infty$ than by the unitary group $U(t)$ of (16) and (14), because an obvious requirement of causality is that:

A state ϕ must be prepared before the observable $|\psi(t)\rangle\langle\psi(t)|$ can be measured in it. The detector cannot count the decay products before the decaying state has been prepared.

This means we have a Quantum Mechanical Arrow of Time:

The Born probability to measure the observable $|\psi(t)\rangle\langle\psi(t)|$ in the state ϕ

$$\mathscr{P}_\phi(\psi(t)) = |\langle\psi(t) \mid \phi\rangle|^2 = |\langle e^{iHt}\psi \mid \phi\rangle|^2 = |\langle\psi \mid e^{-iHt}\phi\rangle|^2 = |\langle\psi \mid \phi(t)\rangle|^2 \quad (17)$$

exists (experimentally) only for $t \geq t_0 (= 0)$.

Here t_0 is the preparation time of the state ϕ.

The physical meaning of time t_0 needs some explanation, so that one does not think of t_0 as a particular time in the life of the experimentalist. Quantum mechanics makes statements about ensembles of microphysical systems. The time t_0 represents thus an ensemble of times $t_0^1, t_0^2, \ldots, t_0^n, \ldots$ at which the 1-st, 2-nd,..., n-th,... microphysical system is prepared in the state ϕ^G.

After one has accepted the preparation time t_0 as the ensemble of times $\{t_0^1, t_0^2, \ldots, t_0^n\}$ at which the n individual micro-systems have been prepared, we choose this time $t_0 = \{t_0^1, t_0^2, \ldots, t_0^n\}$ as the time $t = 0$ for the time evolution of a prepared state, e.g. the decaying state vector $\phi^G(t)$ of an ensemble of decaying micro-system. The task is then to find a new theory (new boundary conditions replacing Hilbert space axiom (13)). This means we need to find new mathematical spaces

> for which the solutions of the Schrödinger equation (12c), $\phi(t) = \mathscr{U}^{\times}(t)\phi$, are given by the *semigroup* $\mathscr{U}^{\times}(t) = \mathrm{e}^{-iHt/\hbar}$, $0 \leq t < \infty$ (18S)

and

> for which the solutions of the Heisenberg equation, $\psi(t) = \mathscr{U}(t)\psi$, are given by the *semigroup* $\mathscr{U}(t) = \mathrm{e}^{iHt/\hbar}$, $0 \leq t < \infty$. (18H)

The Mathematics of Dirac's Theory

The first step in that direction has been taken by Schwartz, using for the Dirac formalism of quantum mechanics one Schwartz space Φ

1. The solutions of *both* the Heisenberg and the Schrödinger equations (for state and observable, respectively) have a Dirac basis vector expansion[4],

$$\phi = \sum_{j,j_3,\eta} \int dE\, |E, j, j_3, \eta\rangle\langle E, j, j_3, \eta \mid \phi\rangle, \tag{19}$$

   where the basis vectors $|E\rangle = |E, j, j_3, \eta\rangle$ are "eigenkets" of H, $\mathbf{J}^2$, J_3 (angular momentum) and other operator η^{op}:

$$\begin{aligned} &\langle \phi \mid H \mid E, j, j_3, \eta\rangle = E\langle \phi \mid E, j, j_3, \eta\rangle \quad \text{for all vectors } \phi \in \Phi. \\ &\mathbf{J}^2|E, j, j_3, \eta\rangle = j(j+1)|E, j, j_3, \eta\rangle, \;\; J_3|E, j, j_3, \eta\rangle = j_3|E, j, j_3, \eta\rangle \\ &\text{and } \eta^{\mathrm{op}}|E, j, j_3, \eta\rangle = \eta|E, j, j_3, \eta\rangle \end{aligned} \tag{20}$$

   To each vector ϕ corresponds precisely *one* function (not a class of functions as for $L^2(E)$).

$$\langle E, j, j_3, \eta \mid \phi\rangle = \phi_{j,j_3,\eta}(E) = \langle E \mid \phi\rangle = \phi(E) \tag{21}$$

2. The co-ordinates or bra-ket $\langle E \mid \phi\rangle = \phi(E)$ are smooth, rapidly decreasing functions of E ("Schwartz function" $\in \mathscr{S}_{\mathbb{R}_+}$)

   One has a triplet of function spaces $\{\phi(E)\} = \mathscr{S} \subset L^2 \subset \mathscr{S}^{\times}$
   and a corresponding triplet of vector spaces $\{\phi\} = \Phi \subset \mathscr{H} \subset \Phi^{\times}$ (22)

   called a Gel'fand Triplet or Rigged Hilbert Space (RHS).

   The Dirac kets $|E\rangle$ are antilinear continuous Schwartz space functionals $|E\rangle \in \Phi^{\times}$.
   The Dirac basis vector expansion (19) is the nuclear spectral theorem in mathematics.

---

[4] E is the continuous analogue of i in $\vec{x} = \sum_{i=1}^{3} \vec{e}_i x^i$, and $\langle E, j, j_3, \eta|\phi\rangle = \phi(E)$ is the continuously infinite analogue of the component $x^i = x(i)$.

Thus in Dirac's formulation there is one RHS, and it is the same RHS for the states ϕ and the observables $|\psi\rangle\langle\psi|$:

$$\{\phi\} = \{\psi\} = \Phi = \text{abstract Schwartz space} \tag{23}$$

In a theory that distinguishes between prepared states $\{\phi\}$ and detected observables $\{\psi\}$ as indicated by (18S) and (18H) one needs two Rigged Hilbert spaces.

The Mathematics of Time Asymmetric Quantum Mechanics

To find a time asymmetric quantum theory that incorporates causality as expressed by (17) we start with the

Phenomenological Theory of Scattering and Decay:

One uses in- and out- plane wave "state" $|E^+\rangle$ and $|E^-\rangle$ which fulfill the Lippmann-Schwinger equation [11]

$$|E^{\pm}\rangle = |E \pm i\varepsilon\rangle = |E\rangle + \frac{1}{E - H \pm i\varepsilon} V|E\rangle = \Omega^{\pm}|E\rangle, \quad \varepsilon \to +0 \tag{24}$$

One speaks of complex energy:

for the analytic S matrix	$S_j(E) \to S_j(z)$
for the Gamow states ϕ^{G}	$z_R = E_R - i\Gamma/2$
for the Lippmann-Schwinger equation or in the propagator of field theory	$z = E \pm i\varepsilon$, ε infinitesimal

The $|E^+\rangle$ are taken as basis systems for the Dirac basis vector expansion of in-state vectors

$$\phi^+ = \sum_{j,j_3,\eta} \int_0^\infty dE |E, j, j_3, \eta^+\rangle\langle^+E, j, j_3, \eta|\phi^+\rangle, \tag{25}$$

and the $|E^-\rangle$ are taken as basis systems for the out-vectors

$$\psi^- = \sum_{j,j_3,\eta} \int_0^\infty dE |E, j, j_3, \eta^-\rangle\langle^-E, j, j_3, \eta|\psi^-\rangle. \tag{26}$$

The Dirac basis vector expansions (25), (26) use *two different* kinds of kets:

$|Ejj_3\eta^{\mp}\rangle \in \Phi_{\pm}^{\times}$ suggested by the Lippmann-Schwinger out-plane waves $|E^-\rangle$ and in-plane waves $|E^+\rangle$, respectively.

Because of the $+i\varepsilon$ in the Lippmann-Schwinger equation, the energy wave function of the prepared in-state ϕ^+:

$$\phi^+(E) \equiv \langle^+Ejj_3\eta|\phi^+\rangle = \overline{\langle\phi^+|Ejj_3\eta^+\rangle} \tag{27+}$$

is the boundary value of an analytic function in the lower complex energy semi-plane (for complex energy $z = \overline{E + i\varepsilon} = E - i\varepsilon$ immediately below the real axis on the second sheet of the S-matrix $S_j(z)$). Similarly, one surmises that the energy wave function of the observable $|\psi^-\rangle\langle\psi^-|$

$$\psi^-(E) \equiv \langle^-Ejj_3\eta|\psi^-\rangle \tag{27+}$$

can be extended into an analytic function in the upper complex energy semi-plane. Thus one conjectures:

Two sets of functions	two sets of vectors	two spaces	from the two L-Sch. kets	
$\{\phi^+(E) = \langle^+E \mid \phi^+\rangle\}$	$\{\phi^+\}$	$= \Phi_-$	$\|E^+\rangle = \|E, j, j_3, \eta^+\rangle$	(28)
$\{\psi^-(E) = \langle^-E \mid \psi^-\rangle\}$	$\{\psi^-\}$	$= \Phi_+$	$\|E^-\rangle = \|E, j, j_3, \eta^-\rangle$	

The spaces Φ_- and Φ_+ will turn out to be well known mathematical spaces, the Hardy spaces (note the change in label for the mathematical spaces and the Lippmann-Schwinger kets).

From the phenomenological S-matrix theory one surmises that the function $\langle\psi^-|E^-\rangle\langle E^+|\phi^+\rangle S_j(E)$ in the integral of the S-matrix element in (29a) and (29b) can be analytically continued into the lower complex semi-plane (second sheet), except for singularities of $S_j(E)$; as shown in Fig. 3.

This means the contour of integration for the S-matrix element

$$\begin{aligned}(\psi^{\text{out}}, S\phi^{\text{in}}) &= (\psi^-, \phi^+) \\ &= \sum_j \int_{E_0}^{\infty} dE \sum_{j_3} \sum_{\eta,\eta'} \langle\psi^-|E, j, j_3, \eta'^-\rangle S_j^{\eta'\eta}(E) \langle^+E, j, j_3, \eta|\phi^+\rangle, \end{aligned} \tag{29a}$$

or in simplified notation:

$$(\psi^-, \phi^+) = \int_{E_0}^{\infty} dE \langle\psi^-|E^-\rangle S_j(E) \langle^+E|\phi^+\rangle, \tag{29b}$$

can be deformed, from the positive real axis first sheet = positive real axis second sheet, into the lower complex energy plane second sheet, where the resonance pole of the S-matrix is located (e.g. at z_R in Fig. 3).

Conjecturing the Hardy Space Axiom

Therefore the program to determine the property of the spaces Φ_- (for states) and Φ_+ (for observables) is the following:
Start with pole of S-matrix at z_R and conjecture the mathematical property of $\langle^-E \mid \psi^-\rangle$, $\langle^+E \mid \psi^+\rangle$ such that a Scattering Resonance *and* a Decaying State can be derived from the S-matrix pole at z_R [12].

Consider the simplest case: there is one S-matrix pole at $z_R = E_R - i\Gamma/2$ on the second sheet, as depicted in Fig. 3. Starting from the integral along the positive real energy axis in (29) one deforms the contour of integration into the second sheet. This leads to the integral along the infinite semicircle C_∞ which is zero for any reasonable assumption for the functions $\langle\psi^- \mid E^-\rangle = \overline{\psi^-(E)}$ and $\langle E^+ \mid \phi^+\rangle = \phi^+(E)$. Then one remains with the

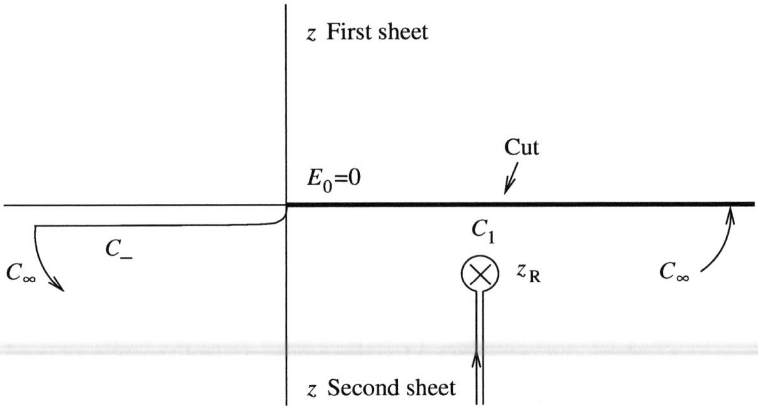

FIGURE 3. The contours of integration for the S-matrix element.

integral C_- in the second sheet, which will describe some background continuum related to $B(E)$ in (6R) (which we shall not discuss here). And there is the integral around the pole position z_R (or the sum of these pole terms if there are more than one S-matrix pole).

We start with this first order S-matrix pole at z_R as the definition of a resonance. From the pole term around $z_R = E_R - i\Gamma/2$ we derive

1. a Breit-Wigner resonance amplitude

$$a_j^{BW}(E) = \frac{R}{E - z_R}, \qquad -\infty < E < +\infty \tag{30a}$$

2. a Gamow state vector defined with this Breit-Wigner amplitude

$$|z_R, j, j_3, \eta^-\rangle = \int_{-\infty}^{+\infty} dE\, |E, j, j_3, \eta^-\rangle \frac{1}{E - z_R} \tag{30b}$$

for which one shows that

1. $|z_R^-\rangle$ is an eigenket of H with a discrete complex eigenvalue (as Gamow wanted)

$$H^\times |E_R - i\Gamma/2^-\rangle = (E_R - i\Gamma/2)\, |E_R - i\Gamma/2^-\rangle \tag{31a}$$

2. it has the property

$$\begin{aligned}\langle e^{iHt/\hbar}\psi_\eta^- |E_R - i\Gamma/2^-\rangle &= \langle \psi_\eta^- | e^{-iH^\times t/\hbar} |E_R - i\Gamma/2^-\rangle \\ &= e^{-iE_R t/\hbar} e^{-(\Gamma/2)t/\hbar} \langle \psi_\eta^- |E_R - i\Gamma/2^-\rangle,\end{aligned} \tag{31b}$$

but this property can be derived only for

$$t \geq t_0\ (=0). \tag{31c}$$

The result (31b) shows that the resonance of width Γ given by the pole term (30a) is represented by an eigenket (30b) of H with complex eigenvalue (31a), which according to (31b) represents a decaying state with lifetime $\tau = \frac{\hbar}{\Gamma}$.

Summarizing: the pole of the S-matrix can be mathematically represented by a Breit-Wigner amplitude (30a) and by a Gamow vector (31b) which is an exponentially decaying state (31b) with a lifetime:

$$\tau = \frac{\hbar}{\Gamma}. \tag{32}$$

In order to derive these results (30) and (31), specific mathematical assumptions must be made about the mathematical properties of energy wave functions $\langle {}^{+}z \mid \phi^{+}\rangle$ and $\langle {}^{-}z \mid \psi^{-}\rangle$ that appear in the integral (29) for the S-matrix element. These conditions lead to the new hypothesis [12]:

The energy wave functions of a state are smooth Hardy functions analytic on the lower complex plane $\mathbb{C}_{-}$:

$$\phi^{+}(E) = \langle {}^{+}E \mid \phi^{+}\rangle \in (\mathscr{H}^{2}_{-} \cap \mathscr{S})_{\mathbb{R}_{+}} = \text{Hardy functions } \mathbb{C}_{-}. \tag{33$_{-}$}$$

The energy wave functions of an observable are smooth Hardy functions analytic on the upper complex plane $\mathbb{C}_{+}$:[5]

$$\psi^{-}(E) = \langle {}^{-}E \mid \psi^{-}\rangle \in (\mathscr{H}^{2}_{+} \cap \mathscr{S})_{\mathbb{R}_{+}} = \text{Hardy functions } \mathbb{C}_{+}, \tag{33$_{+}$}$$

where $\mathbb{C}_{-}$ ($\mathbb{C}_{+}$) denotes the lower (upper) complex energy semi-plane of the S-matrix $S_j(E)$ in (29).

Therewith we have inferred from heuristic results a new mathematical **axiom** of causal quantum theory which applies to resonances and decay:

the Hardy space axiom

The set of prepared (in-) states defined by the preparation apparatus (e.g., accelerator) is $\{\phi^{+}\} = \Phi_{-} \subset \mathscr{H} \subset \Phi_{-}^{\times}$ (34$_{-}$)

The set of (out-) observables defined by the registration apparatus (e.g. detector) is $\{\psi^{-}\} = \Phi_{+} \subset \mathscr{H} \subset \Phi_{+}^{\times}$ (34$_{+}$)

where $\Phi_{\mp}$ are Hardy spaces of the semiplanes $\mathbb{C}_{\mp}$ and the spaces $\Phi_{\mp}^{\times}$ are the duals (i.e., the spaces of antilinear continuous Hardy space functionals).

This new axiom gives a mathematical meaning to Lippmann-Schwinger kets $|E^{\pm}\rangle$ and to the Gamow kets $|E_R - i\Gamma/2^{-}\rangle$ as functionals on the Hardy spaces $\Phi_{\mp}$. And it provides a consistent mathematical theory which unifies resonance and decay phenomena, and the resonance state vector (30b) obeys the exact exponential decay law (31b).

Accepting the new Hardy space axiom (33$_{\pm}$) or (34$_{\pm}$), the dynamical equations (12a-c) are now solved under the Hardy space boundary conditions (34$\pm$) and *not* the Hilbert space boundary condition like in (13a). From the Hardy space axiom does not follow the unitary group evolution (14) which was a consequence of the Stone-von Neumann theorem of the Hilbert space. But in Hardy space a similar theorem holds, the Paley-Wiener theorem. By the Paley-Wiener theorem the solution of the same dynamical equations (12a), (12b), (12c) are given instead by the semigroup solutions (18S) for the state

[5] This is because their complex conjugates $\overline{\psi^{-}(E)} = \langle \psi^{-}|E^{-}\rangle$ are Hardy on $\mathbb{C}_{-}$.

vector $\phi^+(t)$ and (18H) for the observable vector $\psi^-(t)$. This means the mathematical consequence of the Hardy space axiom ($33_\pm$) agrees with the phenomenological condition (17) expressing causality and avoiding the "exponential catastrophe" (15).

The property $t \geq 0$ in (31b) is thus also a mathematical consequence of the Hardy space axiom (33_+) for the Gamow ket, $|z_R, j, j_3, \eta^-\rangle \in \Phi_+^\times$ [6].

From the result (31b) follows that a resonance of width Γ (30a) is an exponentially decaying state with lifetime $\tau = \hbar/\Gamma$; this is an exact relation.

If one observes in experiments "deviations from the exponential law," then it is due to the fact that one has measured not the time evolution of the Gamow state $\phi_j^G = |E_R - i\Gamma/2^-\rangle$ of (30), (31a), (31b), but the time evolution of a superposition (or of a mixture), $\phi_j^G + \phi_j^B$, of ϕ_j^G the exponentially decaying state and some background ϕ_j^B, where ϕ_j^B is connected with a background amplitude $B(E)$ in (6R), which theoretically is contained in the integral along C_- in Fig. 3. The exponential time dependence displayed in Fig. 2 has been obtained in [6] only after the K_L^0 interference and other contributions have been eliminated from the fit in Fig. 2.

The Gamow ket $|z_R^-\rangle$, like the Dirac ket, is a mathematical representation of an idealized state. With a macroscopic apparatus one cannot detect a single energy value out of an energy continuum; this does not diminish the importance of these "singular" mathematical objects, like Dirac kets or Gamow kets, for the understanding of quantum physics.

THE EXPONENTIAL DECAY OF RELATIVISTIC RESONANCES AND THE DEFINITION OF *Z*-BOSON MASS

For the relativistic resonance, specifically the *Z*-boson, one has a multitude of parametrizations of which those of equations (1)–(5) are popular ones. The jth partial scattering amplitude in a relativistic resonance formation process $a_j(s)$ of (4) is a function of the invariant mass square $s = (p_1^\mu + p_2^\mu)^2 = (E_1^{\mathrm{cm}} + E_2^{\mathrm{cm}})^2$, where p_1^μ, p_2^μ are the momenta of the two incoming (or outgoing) particles. One desires to write the amplitude of a resonance scattering process as the sum (4), where $B_j(s)$ is the non-resonant background and a_j^{res} describes the resonance per se. It is not possible to fix the functions $a_j^{\mathrm{res}}(s)$ and $B_j(s)$ separately using only the experimental cross section data for $|a_j(s)|^2$. One needs some theoretical arguments in favor of a particular functional form of a_j^{res}, like, e.g., the Breit-Wigner form (3), or e.g. the on-mass-shell definitions (1a), or the parametrization (2), since only $|a_j(s)|^2$ is experimentally available in scattering experiments, and not $a_j^{\mathrm{res}}(s)$ and $B_j(s)$ separately.

---

[6] The miss-match in the notation for labels $+$ and $-$ for the kets $|z_R^-\rangle$ and spaces $\Phi_+^\times$ has its origin in the physicists notation $|E^\mp\rangle$ for the Lippmann-Schwinger kets and the mathematicians notation for the Hardy function spaces $\mathscr{H}_\pm$ and thus for the abstract spaces: $\Phi_\pm \ni |E_\mp\rangle$. Except for the miss-match in the notation $\pm$, the correspondence between the mathematics of Hardy spaces and the physics of quantum and decay phenomena is a beautiful example of what Wigner called "the miracle of appropriateness of the language of mathematics for the formulation of the laws of physics", E.P. Wigner, *Symmetries and Reflections*, Ch. 17 Ox Bow Press, Woodbridge, Connecticut, 1979.

This means that the selection of the "right" (M,Γ) requires a theory that relates M or Γ to the particle aspects of the relativistic decaying state.

Relativistic particles, stable or decaying, are by hypotheses given by the representations of the inhomogeneous Lorentz transformations [13], as is the standard for stable relativistic particles [5]. Therefore:

Stable relativistic particles are represented by unitary representations of Poincaré group with (mass,spin) given by

$$(m,j) = (\text{real}, \text{semi-integer})$$

Decaying relativistic particles we represent by semigroup representations of $(\text{complex mass}, \text{spin}) = (\sqrt{s_R}, j)$

$$\sqrt{s_R} = (M_R - i\frac{\Gamma_R}{2}) = \sqrt{\overline{M}_Z^2 - i\overline{M}_Z\overline{\Gamma}_Z}. \tag{35}$$

The basis vectors of the semigroup representation $[s_R, j]$ are given, in very close analogy to the non-relativistic Gamow vectors (30b) by

$$|[j,s_R],\hat{\mathbf{p}},j_3^-\rangle = \frac{1}{2\pi}\int_{-\infty_{II}}^{\infty} ds|[j,s],\hat{\mathbf{p}},j_3^-\rangle\frac{1}{s-s_R}, \tag{36}$$

where

$$\hat{\mathbf{p}} = \frac{\mathbf{p}}{s} \equiv \gamma\mathbf{v}; \quad \hat{p}^0 \equiv \gamma = (\sqrt{1-\mathbf{v}^2})^{-1}$$

are the components of the 4-velocity.

The Gamow vectors (36) have been defined in complete analogy to the non-relativistic Gamow vectors (30b) except that in place of E, the relativistically invariant energy square s has been used. In the same way as for the non-relativistic case (31a) one can show that the $|[j,s_R],\hat{\mathbf{p}},j_3^-\rangle$ are generalized eigenvectors of the total invariant mass square operator $P^\mu P_\mu$

$$P^\mu P_\mu|[j,s_R],\hat{\mathbf{p}},j_3^-\rangle = s_R|[j,s_R],\hat{\mathbf{p}},j_3^-\rangle. \tag{37}$$

The value s_R in (36) is the position of the resonance pole of the analytically continued S-matrix, the relativistic analogue of z_R in Fig. 3, which is on the second sheet of the Riemann surface of the S-matrix element $S_j(s)$. The integration in (36) extends over the real energy axis on the second sheet, which coincides with the physical values on the lower edge of the first sheet for $s \geq m_0^2$. This is as shown in Fig. 3, except that the complex energy z has to be replaced by the complex invariant mass squared $s = p^\mu p_\mu = (p_1+p_2)^\mu(p_1+p_2)_\mu$.

As mentioned in the Introduction, for the relativistic resonance, specifically for the Z-boson, one has a multitude of parametrizations of which those of Eqs. (1)– (3) are the popular ones. Fitting $|a_j(s)|^2$ of (4) to the experimental data, like in Fig. 1, cannot discriminate between the various choices (1a), (2) and (3), etc., because one can shift the difference between, e.g., (1a) and (3) into the unknown background $B_j(s)$ of (4). The selection of the "right" parameters (M,Γ), i.e., the choice of the amplitude, which one associates to the resonance per se, requires a theory that relates M and Γ to the particle aspect of the relativistic decaying state. This is given by the irreducible representations of the inhomogeneous Lorentz transformations [13], as a generalization of Wigner's choice of the unitary representations for stable relativistic particles [5].

To calculate the time evolution of these basis vectors (36) for the semigroup representation one proceeds in the following way:
Go to the rest frame $\hat{\mathbf{p}} = (1,0,0,0)$, $\mathbf{v} = 0$. Then (37) becomes:

$$H^{\times}|[j,s_R],\hat{\mathbf{p}}=0,j_3^-\rangle = \sqrt{s_R}|[j,s_R],\hat{\mathbf{p}}=0,j_3^-\rangle, \quad H^{\times}=P_0. \tag{37'}$$

The relativistic state at time t is then obtained by the time evolution in the rest frame; $(\hbar = 1)$:

$$\begin{aligned}\psi^G_{s_R}(t) = \mathrm{e}^{-iH^{\times}t}|[s_R,j]\hat{\mathbf{p}}=0,j_3^-\rangle &= \mathrm{e}^{-i\sqrt{s_R}t}|[s_R,j]\hat{\mathbf{p}}=0,j_3^-\rangle \\ &= \mathrm{e}^{-iM_Rt}\mathrm{e}^{-(\Gamma_R/2)t}|[s_R,j]\hat{\mathbf{p}}=0,j_3^-\rangle, \quad t\geq 0. \end{aligned}\tag{38}$$

From this one obtains, for instance, that the "survival probability at a time $t \geq 0$ is proportional to

$$|\langle\psi^G \mid \psi^G_{s_R,j}(t)\rangle|^2 = \mathrm{e}^{-\Gamma_Rt}|\langle\psi^G \mid \psi^G_{s_R,j}(0)\rangle|^2, \quad t\geq 0 \quad \text{only}. \tag{39}$$

This means that the lifetime τ of the exponential decay is given by

$$\tau = \frac{1}{\Gamma_R}. \tag{40}$$

Only Γ_R of (3), and not $\overline{\Gamma}_Z$, nor Γ_Z of (1a) will be the inverse of an exponential lifetime.

One can start with the definition of the (first order) resonance by the pole of the j-th partial S-matrix element $S_j(s)$, cf., (29) and obtain the relativistic Gamow vector in analogy to the non-relativistic case (30b) as the vector (36) with "relativistic Breit-Wigner" line shape $\frac{1}{s-s_R}$ (obtained from the integral around circle C_1 of the pole at s_R by contour deformation). Then one obtains the exponential time evolution (38) and (39) – which is the relativistic version of (31b) – the lifetime for the Gamow vector (36) is then derived as (40).

The semigroup time evolutions, $t \geq 0$ in (38) for the relativistic Gamow states at rest $|[j,s_R],\hat{\mathbf{p}}=0,j_3^-\rangle$ are just the subset of the *causal* transformations of the relativistic space time. The set of all causal transformations is given by the Poincaré semigroup into the forward light cone,

$$\mathscr{P}_+ = \{(\Lambda,x) : x^2 = t^2 - \mathbf{x}^2 \geq 0, t\geq 0\}. \tag{41}$$

To summarize, of the many possible choices for the definition of mass M for a relativistic resonance, of which (M_Z,Γ_Z), $(\overline{M}_Z,\overline{\Gamma}_Z)$ and (M_R,Γ_R) in (1)– (3) are some examples, only the parametrization $s_R = (M_R - i\Gamma_R)^2$ for the pole position of the S-matrix pole, unifies the various concepts and heuristic notions that have been used successfully in the phenomenological and mathematical description of relativistic resonances and relativistic decaying states: it unifies the Breit-Wigner resonance (36) with lineshape (3) and the exponentially decaying Gamow state (36) with (39), (40). And it relates both to the resonance pole of the S-matrix. The relativistic version of causal time evolution (38), (39) is a special case of "Einstein causality" (41), which states that causal symmetry transformations of relativistic space time are given by the Poincaré semigroup transformation (41).

The lifetime-width relation

$$\tau = \frac{\hbar}{\Gamma} \tag{42}$$

is an exact relation, in the non-relativistic as well as in the relativistic case.

All this is accomplished by a single modification of one mathematical axiom of quantum theory. Instead of choosing for the energy functions (21) the Hilbert space or the Schwartz space wave functions (22), one needs to use the Hardy space axiom ($34_\pm$) and choose the analytic Hardy wave functions, which distinguish mathematically between states (34_-) and observables (34_+).

The Hardy space theory can be applied specifically to the question: What should be used for the mass and the width of the Z-boson, and similarly for the mass and width of any relativistic resonance? The theory, which is based on not much more than the condition of causality, which stated that the Born probabilities (17) need to exist only after the state was prepared, suggests that (M_R, Γ_R) is the right definition of the mass and width for a relativistic resonance. Specifically for the Z-boson it means that

$$M_R = \mathrm{Re}\sqrt{s_R} = 91.1611 \pm 0.0023\ \mathrm{GeV} = M_Z - 0.026\ \mathrm{GeV},$$
$$\Gamma_R = -2\,\mathrm{Im}\sqrt{s_R} = 2.4943 \pm 0.0024\ \mathrm{GeV}$$

are the right values of mass and width of the Z-boson, not $(\overline{M}_Z, \overline{\Gamma}_Z)$ or (M_Z, Γ_Z).

It is therefore difficult to appreciate that the more recent editions of the particle data table [3] quote exclusively the values for (M_Z, Γ_Z) of (1a), which is specific to the on-the-mass-shell-renormalization scheme, especially since it is known now that this choice is not gauge invariant.

These were precisely the questions that concerned Augusto in his scientific work. Starting with his detailed calculations for the hyperon semileptonic decay data in his dissertation at the University of Chicago, in close collaboration with the experimental colleagues, his principal interest was to bring experiment and theory together. Therefore it was quite natural, that he was agitated about experimental masses, which violated the theoretical principles of gauge invariance, and stirred me up about it.

Augusto has been for many years a valuable collaborator and a dear friend and adviser, not only on professional matters, but also on personal problems; wise, intelligent and helpful, he was a joy to be with. A passionate music lover, particularly of Richard Strauss and Mozart, Augusto with his authoritative bass voice, could discuss the intricacies of Chicago politics and yet, within the next sentence, reveal his depth of understanding of the human condition.

ACKNOWLEDGMENTS

The author would like to thank Miguel A. Perez and the administration of CINVESTAV for providing this opportunity to express our appreciation of Augusto Garcia for his contributions to physics and to the enrichment of our lives.

He expresses his gratitude to P. Kielanowski and P. Bryant for their help with this manuscript. He also gratefully acknowledges partial support by the US NSF (OISE-0421936).

REFERENCES

1. S. Eidelman et al. (Particle Data Group), *Phys. Lett.* **B592**, 1 (2004).
2. A. Sirlin, *Phys. Rev. Lett.* **67**, 2127–2130 (1991); A. Sirlin, *Phys. Lett.* **B267**, 240–242 (1991); S. Willenbrock, and G. Valencia, *Phys. Lett.* **B259**, 373–376 (1991); R. Stuart, *Phys. Lett.* **B262**, 113–119 (1991); R. Stuart, *Phys. Lett.* **B272**, 353–358 (1991); A. Leike, T. Riemann, and J. Rose, *Phys. Lett.* **B273**, 513–518 (1991).
3. C. Amsler et al. (Particle Data Group), *Phys. Lett.* **B667**, 1 (2008).
4. A. Bernicha, G. López Castro, and J. Pestieau, *Nucl. Phys. A* **597**, 623–635 (1996); A. Bernicha, G. López Castro, and J. Pestieau, *Phys. Rev. D* **50**, 4454–4461 (1994).
5. S. Weinberg, *The Quantum Theory of Fields*, vol. I Foundations, Cambridge University Press, 1995.
6. L. K. Gibbons et al., *Phys. Rev. D* **55**, 6625–6715 (1997); L. K. Gibbons, *A precise measurement of the CP violation parameter* $\mathrm{Re}(\varepsilon'/\varepsilon)$ *and other kaon decay parameters*, Ph.D. thesis, Univ. of Chicago (1993); L. K. Gibbons et al., *Phys. Rev. Lett.* **70**, 1199–1202 (1993).
7. V. Weisskopf, and E. Wigner, *Z. Phys.* **63**, 54 (1930); V. Weisskopf, and E. Wigner, *Z. Phys.* **65**, 18 (1930).
8. M. Goldberger, and K. Watson, *Collision Theory*, Wiley, New York, 1964, chap. 8.
9. G. Gamow, *Z. Phys.* **51**, 204 (1928); cf. also J.J. Thomson, *Proc. London Math. Soc.* **15**, 197 (1884).
10. J. von Neumann, *Mathematical Foundations of Quantum Mechanics*, Springer Verlag, Berlin, 1932, Princeton University Press, 1995.
11. B. Lippmann, and J. Schwinger, *Phys. Rev.* **79**, 469 (1950); M. Gell-Mann, and H. Goldberger, *Phys. Rev.* **91**, 398 (1953); R. Haag, *Fortschr. Phys.* **7**, 183 (1959); R. Newton, *Scattering Theory of Waves and Particles*, Springer-Verlag, New York, 1982, 2nd ed., chap. 7.
12. For the non-relativistic case this derivation is given, e.g., in A. Bohm, *Quantum Mechanics: Foundations and Applications*, 3-rd Edition, Springer-Verlag, New York (1979), Section XXI.4.
13. For the relativistic case see A. Bohm, H. Kaldass, S. Wickramasekara, *Fortschr. Phys.* **51**, No. 6 (2003).

An overview to networks and its applications

Rodrigo Huerta-Quintanilla and Christian H. Sanabria M.

Centro de Investigación y de Estudios Avanzados del Instituto Politécnico Nacional Unidad Mérida, Departamento de Física Aplicada.
Km 6 carretera antigua a progreso, Mérida, Yucatán.

Abstract. We present an introduction to the basics on networks and their application to econophysics. In particular we study a model in which agents interact through a network chosen in a very specific way and the exchange they make of a given asset. We study different types of exchange interactions and also the effect of the network on the dynamics.

Keywords: Econophysics, social networks, wealth distribution
PACS: 87.23.Ge, 89.65.-s, 89.65.Gh, 89.75-k, 89.75.Fb

INTRODUCTION TO NETWORKS

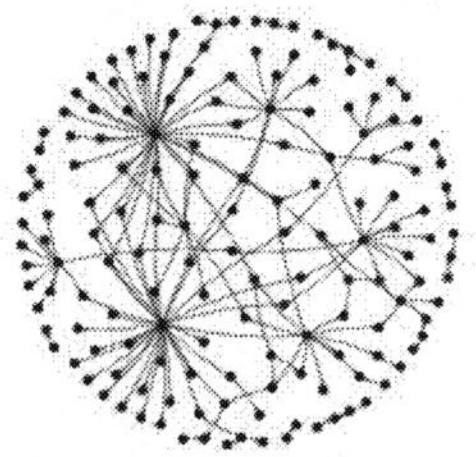

FIGURE 1. Network sketch

A network (see figure 1) $\Gamma = (N, L)$ consists of a set of nodes or vertices $N = 1, 2, 3, \ldots, n$ and links among them $L \subseteq N \times N$. We assume there is only one link between any i, j nodes (i, j) and there are no links with itself $(i, i) \notin L$. One way to represent the network is using an $n \times n$ matrix A, where n is the total number of vertices and to denote a_{ij} in the following way

$$a_{ij} = \begin{cases} 1 & \text{If } (i,j) \in L \\ 0 & \text{Otherwise.} \end{cases}$$

Networks can be *directed* or *undirected*. If they are directed, the existence of a link (i, j) does not imply that the converse link (j, i) should necessarily exist. While if are undirected, the existence of a link (i, j) implies that (j, i) exist.

The whole set of nodes connected to a particular node $i \in N$ define its *neighborhood*, such that,

$$\mathcal{N}^i \equiv \{j \in N : (i, j) \in L\}.$$

CP1259, *First CINVESTAV-UNAM Symposium on High Energy Physics*
edited by J. C. D'Olivo, R. López-Fernández, and M. Á. Pérez

And the cardinality of this neighborhood represent its *connectivity* (or *degree*) $\mathscr{Z}^i$ given by

$$\mathscr{Z}^i \equiv |\mathscr{N}^i| \equiv \{j \in N : (i,j) \in L\}.$$

Then the connectivity of any node, is just the number of its neighbors. This connectivity is to first order, however One can also define connectivity for the second neighbors, third, etc. (see reference [1]).

The degree distribution $p(k)$ is the probability that an node i, chosen randomly, has k neighbors. That is if there are n nodes in a given network, and n_k of them have degree k, then $p(k) = n_k/n$. In a formal way we can write,

$$p(k) = \frac{1}{n}|\{i \in N : \mathscr{Z}^i = k\}|$$

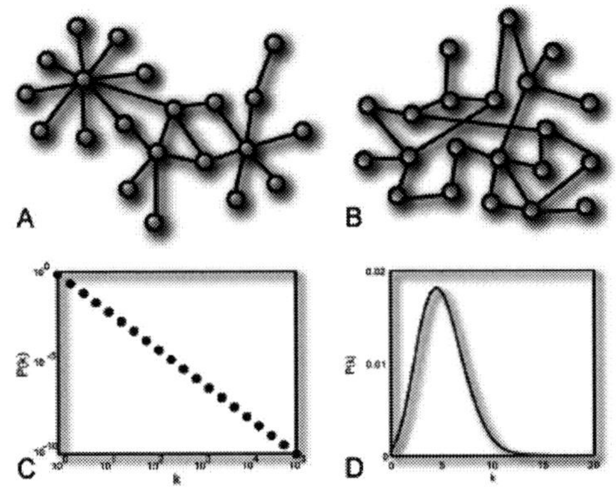

FIGURE 2. (A) Example of a free scale network. (B) Example of a random network. (C) Power law degree distribution. (D) Poisson degree distribution.

Each network is caracterized by its degree distribution. There are many different kinds but the usual ones are random, scale free, small world, regular, etc. Some examples are given in figure 2.

In general a network $\Gamma = (N,L)$ may be divided into different *components*. These are maximal subsets of nodes $\chi \subset N$, such that, for every i, $j \in \chi$, there is a path joining them. That is, there is a (finite) set of links $\{y_{u-1}y_u \in L\}_{u=1,\ldots m}$ such that $y_0 = i$ and $y_m = j$. The *size* of a component is simply the number of nodes that belong to that particular component.

There are some special cases of networks which have a *giant component*, that is a unique component whose relative size remains bounded above zero as the number of nodes goes to infinity..To see when a giant component is present in a network we have to use graph theory to describe its properties.

Graph theory in random networks

We start with the definition of generating function [2],

$$G_0(x) = \sum_{k=0}^{\infty} p(k)x^k, \tag{1}$$

where $p(k)$ is the degree distribution associated to a particular network. As $p(k)$ is normalized, then $G_0(x)$ converges absolutely and does not has singularities in $|x| \leq 1$. $G_0(x)$ is very important since we are going to obtain the parameters we want from it.

We first take the derivative of $G_0(x)$ and evaluate the result in $x = 1$, we have

$$G_0'(1) = \sum_{k=0}^{\infty} kp(k) =< k >\equiv z_1, \tag{2}$$

which is the mean value for the connectivity or the average degree in that distribution. The generating function for the second neighbors is defined as

$$\sum_{k=0}^{\infty} p(k)[G_1(x)]^k = G_0(G_1(x)), \tag{3}$$

with $G_1(x) = \frac{1}{z_1}G_0'(x)$, then the mean value for second neighbors is

$$z_2 \equiv \frac{d}{dx}[G_0(G_1)]_{x=1} = G_0'(1)G_1'(1) = G_0''(1). \tag{4}$$

A giant component will exist in a network if we have the condition $z_2 > z_1$.

APPLICATION TO ECONOPHYSICS

Class formation in a social network with asset exchange

The model we want to describe is one in which agents exchange assets and they are nodes of a network, so they interact only with their neighbors. We have used random networks in which there is a maximum for k, k_{max}, then $p(k)$ is given by.

$$p(k) = \begin{cases} \frac{1}{k_{max}}; & k = 1,2,\ldots k_{max} \\ 0; & \text{otherwise} \end{cases} \tag{5}$$

In this work we have used 4 different values of k_{max} ($k_{max} = 2,3,4$ and 20), hence, we have 4 different networks.

The generating function we obtain is

$$G_0(x) = \frac{x}{k_{max}} \frac{1-x^{k_{max}}}{1-x}. \tag{6}$$

Now from this expression we can obtain values for z_1 and z_2. They are given in table 1.

TABLE 1. Values of z_1 and z_2 for each value of k_{max} used in the work. We can observe that only for $k_{max} = 2$, $z_1 > z_2$. It means there is no a giant component.

k_{max}	z_1	z_2
2	1.5	1
3	2.0	2.66
4	2.5	5.0
20	10.5	133

From table 1 we can realize that only for $k_{max} = 2$ there is not a giant component.

In the following we define the specific rules which agents have in order to interact. The agents are represented as nodes and all of them have the same amount of wealth at time $(t = 0)$, $(w_i(0) = w_0,\ \forall i \in N)$. The links in the network represent the ability to make transactions between agents, where the transaction is wealth exchange.The wealth exchange between two agents, i, j such that $(i, j) \in L$, is the kind of collision type, in which the total amount of money exchange is conserved, represented in the following pair of equations

$$w_i(t+\Delta t) = w_i(t) + \Delta w, \tag{7}$$
$$w_j(t+\Delta t) = w_j(t) - \Delta w. \tag{8}$$

The probability for agent i to win is p, whereas in the same interaction, the probability to win of agent j is $1 - p$. Depending on the different ways we select the value for Δw, we would have different kinds of models. In particular we are going to study two: the additive and the multiplicative exchange, the latter also known as the *yard sale model.*

In the model with additive exchange the money that is exchange in every interaction is fixed and constant ($\Delta w = const$). In the other hand, for the multiplicative exchange, the amount of money exchanged is variable, given by a closer integer fraction of the poorer agent,

$$\Delta w = round(\nu \cdot \min(w_i(t), w_j(t))), \quad 0 < \nu < 1. \tag{9}$$

The simulations were made to study the wealth distribution among the agents and also we use the entropy of the system as a physical parameter. We use the definition given by Shannon,

$$S = -\sum_k P_k \ln P_k, \quad \text{con } P_k = n_k/N. \tag{10}$$

Computational simulation

We have seen in the previous section that for most of the cases, in the network appears a giant component. This condition will give us a different behavior in the network dynamics. In particular we have chosen four values for k_{max}, 2, 3, 4 and 20. The corresponding values given by the simulation for $\langle k \rangle$ are 1.468, 1.952, 2.428 y 10.06 respectively. We have plotted the component size distribution for each network.

The networks was made in the following way: starting from a set of N nodes, where each node is allowed to have a random number of links in the interval $[1, k_{max}]$. After this, every node select randomly one of the $N-1$ nodes and make a link with it, if is possible. Otherwise this node should to select another one, whose link can be done.

The number of nodes used in this simulation was 500. The component size distribution that we obtain are shown in the Figure 3. One can see that the only network with no giant component is the one with $k_{max} = 2$. This condition will allow, for this kind of networks, to present an special behavior due to the fact that some agents are isolated from the rest of the network.

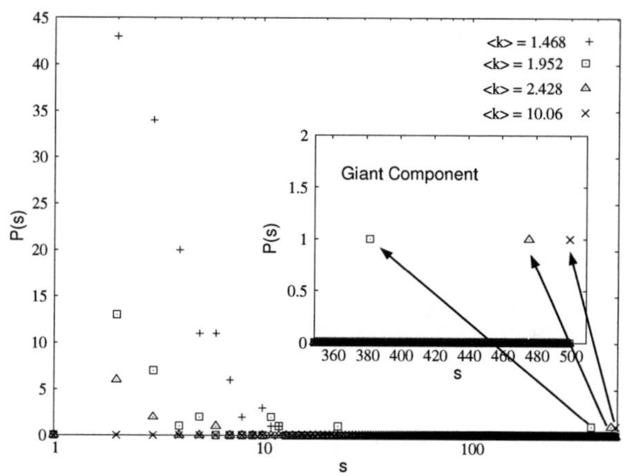

FIGURE 3. Component size distribution for different values of connectivity $\langle k \rangle$. One can see that for $\langle k \rangle = 1.468$ there is no giant component, as opposed to the other three cases (see inset).

Next, with these kind of networks we study the models of wealth exchange just described, additive and multiplicative exchange, to run for a certain period of time, which is normally taking, when the system reach equilibrium. We have chosen the following parameters: Δw equal to 20 for additive case, $\nu = 0.2$ in the multiplicative case. In both cases we made 100 different experiments with $4x10^5$ (additive) and $4x10^6$ (multiplicative) Monte Carlo steps (MCS) respectively.The initial wealth was taken as $\langle w \rangle = w_0 = 100$ units and the probability to win p was 0.5.

Results additive exchange

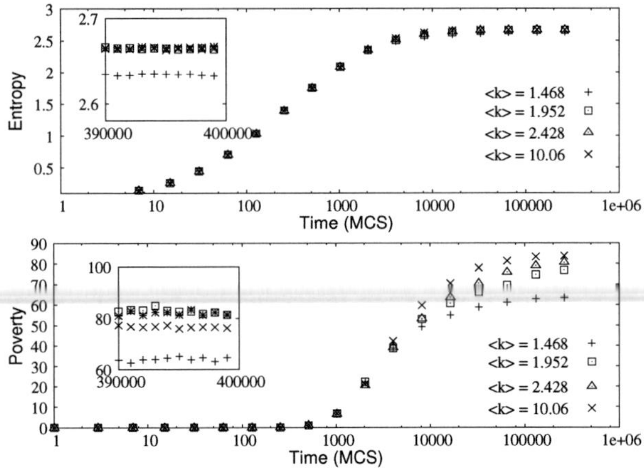

FIGURE 4. Time evolution of entropy (a) and poverty (b) in the additive exchange case for different values of $\langle k \rangle$.

In Figure 4 we show the entropy variation as well as poverty[1] with respect to time. Entropy goes from zero (all agents have the same amount of money) to a maximum value, which corresponds to the stationary limit. (Fig 4(a)). One can see in the inset that the entropy limit is different depending whether it is present a giant component or not. In the case $k_{max} = 2$ there are less available states.

For the same reasons we notice in (Fig 4(b)) that poverty (for $k_{max} = 2$) reaches a different limit depending on k_{max} value.

In Figure 5 we display the wealth distribution for $k_{max} = 2$ when the simulation has driven the system to its stationary state. One can easily see to different regimes separated around the wealth value of $w_c = 2\langle w \rangle$. When $w < w_c$ the distribution is of the Boltzmann-Gibbs type.

$$P(w) = Ce^{-\frac{w}{b\langle w \rangle}}, \tag{11}$$

where the parameter b is equal to 2. In the other hand when $w \geq w_c$, the distribution fits well to a Pareto,

$$P(w) = \frac{C}{x^{1+\alpha}}, \tag{12}$$

where α is equal to 2.5. This particular behavior, the coexistence of two different wealth distributions, has been observed in real economic data (see Figure 6). The data seem

[1] defined as the minimum state of wealth, where the amount of money is less than minimum allowed exchange

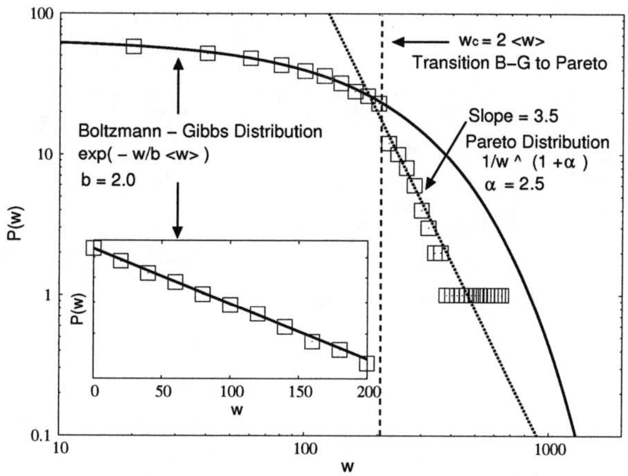

FIGURE 5. Wealth distribution in the additive exchange case, in time step 399000 MCS for $\langle k \rangle = 1.468$. The solid curve fits an exponential function (see inset for a log-normal scale), while the dashed curve fits to a power law function. Two different regimes are clearly seen separated at $2\langle w \rangle$: a Boltzmann-Gibbs distribution for values below to $2\langle w \rangle$, and a Pareto distribution for those above to $2\langle w \rangle$. This behavior is similar to Figure 6.

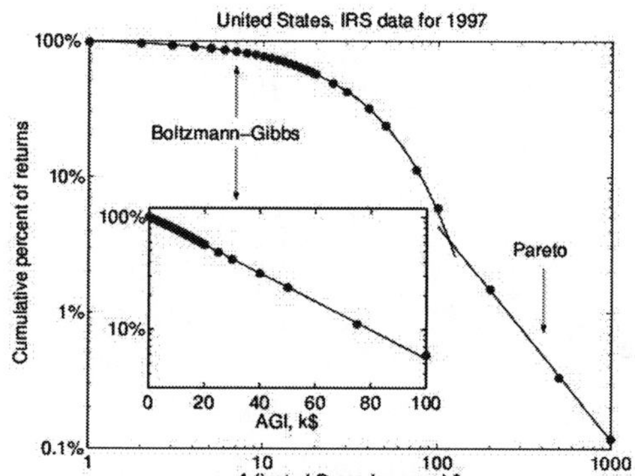

FIGURE 6. Cumulative probability distribution of US individual income for 1997 (fig (9) from **arXiv:cond-mat/0307341v2**).

to indicate that the two regimes are due to the fact that there is no giant component in the network, because the presence of size two components is quite large (see Fig 3) and the formation of isolated components with twice average wealth is an indication of the separation of two different regimes.

In Figure 7 it is shown the wealth distribution for the other set of k_{max} values. As one can see there is no separation in two regimes in this case. The Boltzmann-Gibbs (eq (11)) is the unique distribution and the parameter b changes according to k_{max}, b tend to 1

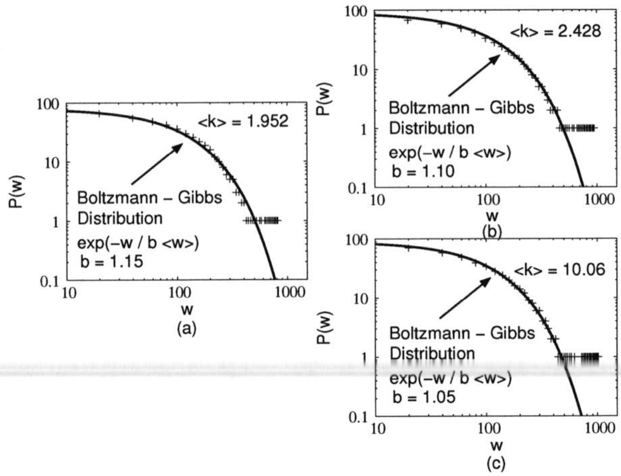

FIGURE 7. Wealth distribution in the additive exchange case, in time step 399000 MCS for different values of $\langle k \rangle$.

(Distribution B-G without network [3, 4]) when k_{max} goes to $N-1$. Table 2 shows this behavior.

TABLE 2. Values of b for different average connectivity $\langle k \rangle$ in the additive exchange case

$\langle k \rangle$	b
1.952	1.15
2.428	1.1
10.06	1.05

Results multiplicative exchange

In Figure 8 we display the variation of entropy and poverty with respect time. In these kind of processes the stationary state is reached after more MCS, because the system goes through a maximum in the entropy before to get the stationary condition. The difference with the previous case is that in this system condensation[2] is present [5]. This is the reason why the entropy decreases for large times and the poverty increases (see Fig 8(b)).

Figure 9 shows the wealth distribution for $k_{max} = 2$. One can see that for the present conditions we obtain 4 different classes: low. medium low, medium high and high (isolated

[2] condensation wealth occur when one agent get the total wealth of the system.

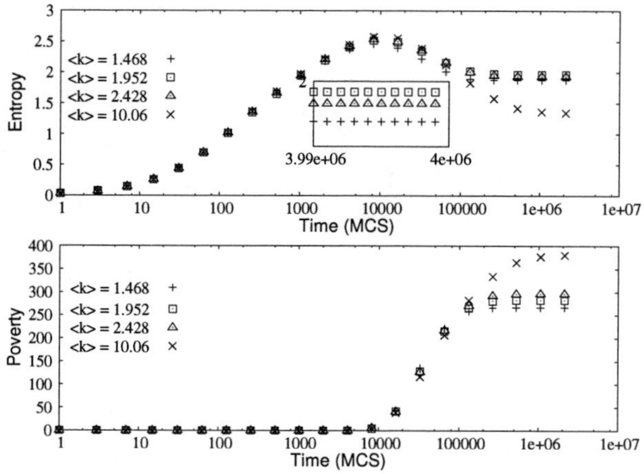

FIGURE 8. Time evolution of entropy (a) and poverty (b) in the multiplicative exchange case for different values of $\langle k \rangle$.

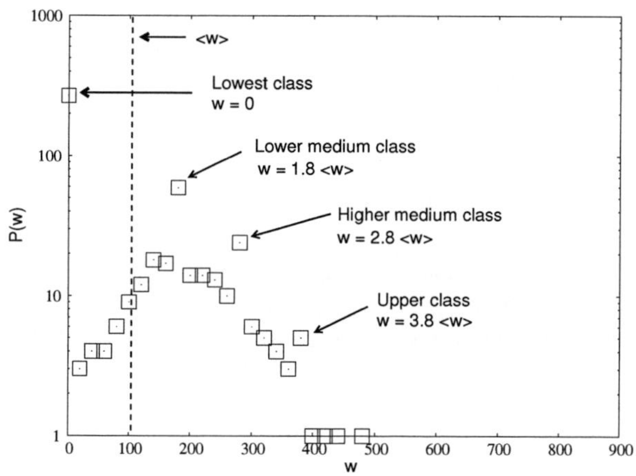

FIGURE 9. Wealth distribution in the multiplicative exchange case, in time step 399000 MCS for an average connectivity value of $\langle k \rangle = 1.468$.

points in the Fig 9). One can explain this pattern analysing the component size distribution for $\langle k \rangle = 1.468$ (Fig 3) since the most common sizes are made of 2, 3 and 4 agents, then these components tend to wealths $2\langle w \rangle$, $3\langle w \rangle$ and $4\langle w \rangle$ respectively. The other components, larger than 4, contribute to the other wealth distribution present in Figure 9.

Figure 10 show the plots of the wealth distributions for k_{max} larger than 2. When k_{max} increases, the classes start to disappear, since the presence of a giant component, that makes the other components to vanish. With the existence of a giant component, the isolated components are rare and the money can travel for almost all system, obtain

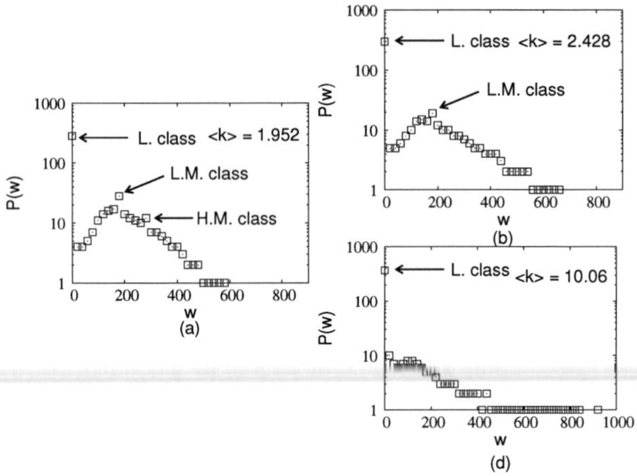

FIGURE 10. Wealth distribution in the multiplicative exchange case, in time step 399000 MCS for different values of $\langle k \rangle$.

from this way an unique wealth distribution (in the case of $k_{max} = 20$ Figure 10(c) the distribution obtained is a double peak distribution).

SUMMARY AND CONCLUSIONS

The use of networks in the field of econophysics and others, opens new doors for research. Particularly in this work, the use of networks allowed us to find very interesting results, as is the formation of classes, applied to very simple models of asset exchange.

REFERENCES

1. F. Vega-Redondo, *Complex Social Networks*, Cambridge University Press, 2007.
2. M. E. J. Newman, S. H. Strogatz, and D. J. Watts, *Phys. Rev. E* **64**, 026118–1 – 026118–17 (2001).
3. S. Ispolatov, P. Krapivsky, and S. Redner, *Eur. Phys. J. B* **2**, 267 – 276 (1998).
4. A. Drăgulescu, and V. Yakovenko, *Eur. Phys. J. B* **17**, 723 – 729 (2000).
5. C. Moukarzel, S. Gonçalves, J. Iglesias, M. Rodríguez-Achach, and R. Huerta-Quintanilla, *Eur. Phys. J. Special Topics* **143**, 75 – 79 (2007).

Radiative processes: Potentials or Fields?

Matías Moreno* and Gabriel López Castro †

*Instituto de Física, Universidad Nacional Autónoma de México, México, D.F. 04510, México
†Departamento de Física, Centro de Investigación y Estudios Avanzados, México, D.F., México

Abstract. In this work we prove that the transition amplitudes of processes involving the emission of one or more real photons is gauge invariant through terms that are always proportional to the electromagnetic field tensor.

Keywords: Radiative processes, Low energy theorems.
PACS: 12.20.-m, 13.40.Hq, 14.70.Bh

INTRODUCTION

Since his early research works, Augusto García was very concerned with the precision tests of the Cabibbo model in hyperon semileptonic decays (HSD) [1]. This interest led him to the consideration of radiative corrections, which he studied in a comprehensive form for HSD until the end of his fruitful life. As is well known, these hyperon decays have played an important role in testing the SU(3) flavor symmetry and in the determination of basic parameters of the standard model as is the case of the quark mixing angles.

Augusto focused on the calculation of observables associated to HSD which could be extracted from experiment in a model-independent way. In particular, his approach to radiative corrections consisted in the separation of these effects into their model-dependent and model-independent terms. In his view, it was possible to evaluate the model-independent part of radiative corrections by considering essentially the emission and re-absorption of virtual and real photons from the static electromagnetic properties of the particles involved in the decay process. The amplitude thus calculated contains all the ultraviolet divergencies, does not have infrared divergencies and it is gauge-invariant.

Here we are concerned with the gauge invariance of transition amplitudes for processes involving the emission of a real photon. Although gauge invariance can be easily implemented at the lagrangian level via the minimal coupling, some care is usually required at the amplitude level, particularly when (a) several Feynman amplitudes contribute to the same radiative process, (b) when intermediate unstable particles are involved, or (c) when the structure of particles is taken into account. We show that, for the important case of radiative transitions, the total amplitude can be expressed always in a form that is proportional to the electromagnetic field tensor $F_{\mu\nu}$ in momentum space. Since $F_{\mu\nu}$ is invariant under gauge transformations, gauge invariance of the radiative amplitude is automatically satisfied. This is in contrast to the usual way of expressing the amplitude as proportional to ε_μ, the polarization four-vector of the photon, a result that follows from the Feynman rules for an external photon field.

CP1259, *First CINVESTAV-UNAM Symposium on High Energy Physics*
edited by J. C. D'Olivo, R. López-Fernández, and M. Á. Pérez

We will first analyze some particular cases which give a proper perspective of the nature of the problem and afterwards present the general proof.

PARTICULAR CASES

Let us take the radiative leptonic decay of a pseudoscalar meson, which we illustrate with the $\pi(P) \to e(p)\,\nu_e\,\gamma(k)$ process. This was worked out long ago by Neville [2] and Brown and Bludman [3]. The amplitude separates into the inner bremsstrahlung (IB) term which is determined by the photon being emitted form the charged external particles, and the model-dependent (MD) term which takes into account the structure of particles and other possible coupling of photons.

In general the radiation from the external legs must be appended by an extra contact term that makes the IB gauge invariant. For the process at hand this term vanishes. This invariant term defines the model independent (MI) amplitude for radiative processes, according to Low's low energy theorem. The IB (MI) amplitude for this decay is, up to an overall coupling constant,

$$\mathcal{M}_{IB} \cong \bar{u}(l)\left[\frac{p\cdot\varepsilon^*}{p\cdot k} - \frac{P\cdot\varepsilon^*}{P\cdot k} + \frac{i\sigma_{\mu\nu}F^{\mu\nu *}}{4p\cdot k}\right](1+\gamma_5)\mathrm{v}(\nu),$$

and the structure dependent part can be subdivided into the model dependent axial

$$\mathcal{M}_{MDA} =< \gamma|A_\mu(0)|0 > l^\mu \cong ia\widetilde{F}^{\mu\nu *}P_\mu l_\nu\ ,$$

and vector

$$\mathcal{M}_{MDV} =< \gamma|V_\mu(0)|0 > l^\mu \cong ibF^{\mu\nu *}P_\mu l_\nu$$

amplitudes. In the above expressions, $l_\nu = \bar{u}(p)\gamma_\nu(1+\gamma_5)\mathrm{v}(\nu)$ denotes the weak leptonic current and we have defined the field tensors

$$F^{\mu\nu} = \varepsilon^\mu k^\nu - \varepsilon^\nu k^\mu, \qquad \widetilde{F}^{\mu\nu} = \frac{1}{2}\varepsilon^{\mu\nu\alpha\beta}F_{\alpha\beta}.$$

Note that the infrared terms in the IB amplitude can be written as:

$$I(p,P) \equiv \frac{p\cdot\varepsilon^*}{p\cdot k} - \frac{P\cdot\varepsilon^*}{P\cdot k} = \frac{p_\mu P_\nu}{p\cdot kP\cdot k}F^{\mu\nu *}\ . \qquad (1)$$

From the above expressions, we observe that the full amplitude $\mathcal{M} = \mathcal{M}_{IB} + \mathcal{M}_{MDA} + \mathcal{M}_{MDV}$ of the $\pi \to e\nu\gamma$ decay becomes proportional to the field strength tensor, namely $\mathcal{M} = F^{\mu\nu *}\mathcal{M}_{\mu\nu}$. Therefore, this expression for the decay amplitude is explicitly gauge invariant, in contrast to the usual form of the amplitude $\mathcal{M} = \varepsilon^\mu\mathcal{M}_\mu$ which requires that the condition $k^\mu\mathcal{M}_\mu = 0$ be satisfied.

This behavior, however, is not exclusive to the radiative pion leptonic decay amplitude. Indeed, we can easily check that the amplitudes for the radiative decays (see for example [4]) $V^\pm \to P^\pm P^0\gamma$, $V^0 \to P^+P^-\gamma$, or a charged fermion into a vector and a neutrino $\tau^\pm \to V^\pm\nu_\tau\gamma$, where V is a vector and P a pseudoscalar meson, can also be expressed in terms of the electromagnetic field strength tensor.

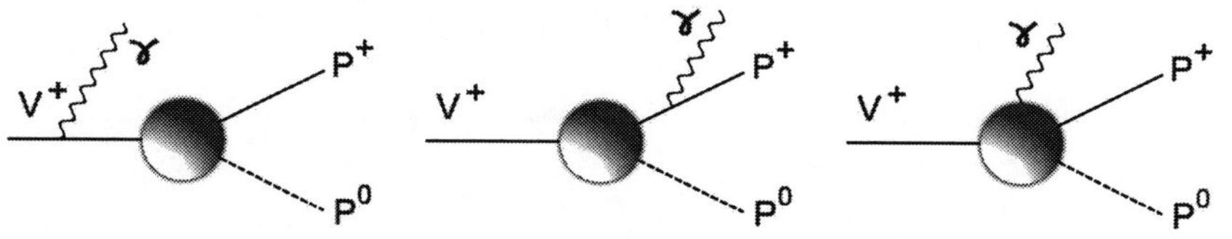

FIGURE 1. Feynman graphs for the radiative vector meson decay $V^{\pm} \to P^{\pm}P^{0}\gamma$. The first two amplitudes are the model independent terms and the third includes model dependent contributions.

Just to illustrate the point, let us consider the decay $\rho^{+}(d,\eta) \to \pi^{+}(p)\pi^{0}(p')\gamma(\varepsilon,k)$ (see Figure 1). Here, η and ε denote the polarization four-vector of the vector meson and photon, respectively, and other letters within parenthesis are the four-momenta of particles. The inner bremsstrahlung amplitude is given by [5]:

$$\begin{aligned}\mathcal{M}_{MI} &= ieg\left\{\left(\frac{p\cdot\varepsilon}{p\cdot k}-\frac{d\cdot\varepsilon}{d\cdot k}\right)(p-p')\cdot\eta \right.\\ &\quad \left. +\left(1-\beta\frac{p\cdot k}{d\cdot k}\right)\left[\frac{p\cdot\varepsilon}{p\cdot k}k\cdot\eta-\varepsilon\cdot\eta\right]+\frac{\beta}{2}\left[\frac{d\cdot\varepsilon}{d\cdot k}k\cdot\eta-\varepsilon\cdot\eta\right]\right\}\\ &= ieg\left\{I(p,d)(p-p')\cdot\eta+k\cdot\eta\left[\left(1-\beta\frac{p\cdot k}{d\cdot k}\right)I(p,\eta)+\frac{\beta}{2}I(d,\eta)\right]\right\}\ . \qquad (2)\end{aligned}$$

In this expression, g denotes the $\rho\pi\pi$ coupling constant, e is the electron charge and β denotes the ρ^{+} meson magnetic dipole moment in units of $e/2m_{\rho}$. Note that the function $I(A,B)$ was introduced in eq. (1).

If we allow for CP-violation, we can get a contribution to the amplitude due to a photon radiated from the ρ^{+} electric quadrupole moment given by [4]:

$$\mathcal{M}_{MI}^{EQM} = -\frac{ieg\gamma(0)}{2}\{(2p-d)\cdot k(k\cdot\eta)I(d,\eta)-2k\cdot\eta(p\cdot k)I(p,d)\}\ , \qquad (3)$$

where $\gamma(0)$ denotes the electric quadrupole moment of the ρ^{+} in units of e/m_{ρ}. Finally, the decay chain contribution $\rho^{+} \to \pi^{+}\omega \to \pi^{+}\pi^{0}\gamma$ gives rise to a model-dependent contribution in the decay amplitude [4]:

$$\mathcal{M}_{MD}^{\omega} \propto -2\eta^{\theta}\widetilde{F}_{\alpha\delta}(d-p)^{\alpha}d^{\lambda}(d-p)^{\tau}\varepsilon_{\lambda\theta\tau\delta}\ . \qquad (4)$$

We can observe that all the contributions to this decay amplitude, eqs. (2)–(4), have been written in a way proportional to the electromagnetic field tensor, thus they are explicitly gauge-invariant. Also, it is interesting to note that the amplitudes that originate from the photon emission off the external lines are proportional to the gauge-invariant function $I(A,B)$. Terms in the amplitude proportional to the dual tensor $\widetilde{F}_{\mu\nu}$ arise only from model-dependent contributions.

GENERAL CASE

Let us consider a general form for the amplitude of a radiative process

$$\mathcal{M} = \varepsilon^{\mu} \mathcal{M}_{\mu} \,. \tag{5}$$

Electromagnetic gauge invariance requires that $k^{\mu} \mathcal{M}_{\mu} = 0$. We shall now prove that it can always be written as:

$$\mathcal{M} = F^{\mu\nu} \mathcal{M}_{\mu\nu}. \tag{6}$$

Since the electromagnetic field tensor is invariant under gauge transformations, this form of the amplitude is explicitly gauge-invariant. In the above expressions, neither $\mathcal{M}_{\mu}$ nor $\mathcal{M}_{\mu\nu}$ depend on the photon polarization four-vector ε. For a real photon, we have of course $\varepsilon \cdot k = k \cdot k = 0$, where we use the covariant gauge $\partial \cdot A = 0$.

It is clear that, in four dimensions, $\mathcal{M}_{\mu}$ can be expanded in terms of any set of four linearly independent directions. It is natural to take two of them as the photon momentum k_{μ} and polarization ε_{μ}; the other independent four vectors will be denoted by a_{μ} and b_{μ}. Then the linear independence of these four-vectors implies that the volume formed by them is non null, explicitly

$$\varepsilon^{\mu\nu\alpha\beta} \varepsilon_{\mu} k_{\nu} a_{\alpha} b_{\beta} \neq 0 \,. \tag{7}$$

Any four vector can be expanded in terms of this basis. Thus

$$\mathcal{M}_{\mu} = a_{\mu} A + b_{\mu} B + \varepsilon_{\mu} C + k_{\mu} D$$

where A, B, C and D are Lorentz-invariant functions whose specific form depends on the particular process.

The coefficient C must vanish owing to the independence of $\mathcal{M}_{\mu}$ upon ε. On the other hand, the conservation of electromagnetic current implies that

$$\begin{aligned} k \cdot \mathcal{M} &= k \cdot a A + k \cdot b B + k^2 D \\ &= 0 \,, \end{aligned} \tag{8}$$

thus, for on-shell photons we have

$$B = -\frac{k \cdot a}{k \cdot b} A \,,$$

We may thus conclude that

$$\mathcal{M}_{\mu} = (a_{\mu} - b_{\mu} \frac{k \cdot a}{k \cdot b}) A + k_{\mu} D$$

Due to the gauge (or more generally to the transversality) condition $\varepsilon \cdot k = 0$ the last term in the previous equation never contributes to the amplitude. Therefore

$$\mathcal{M}_{\mu} = (a_{\mu} - b_{\mu} \frac{k \cdot a}{k \cdot b}) A \tag{9}$$

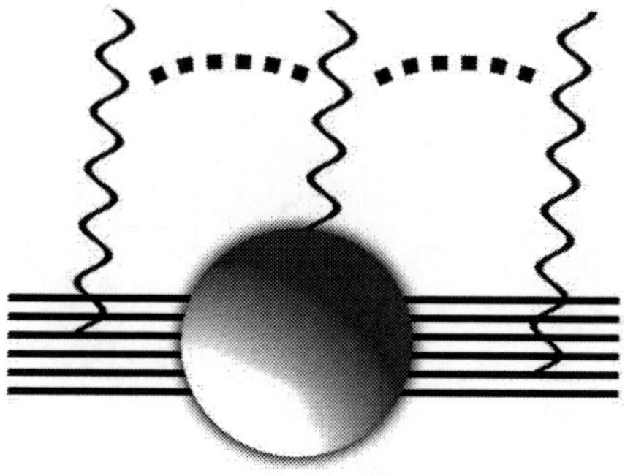

FIGURE 2. Schematic representation of a multiphoton emission amplitude.

or

$$\begin{aligned}\mathcal{M} &= \varepsilon^{\mu}\mathcal{M}_{\mu} \\ &= k\cdot a\, I(a,b)A \\ &= F^{\mu\nu}a_{\mu}b_{\nu}\tilde{A}\ , \end{aligned} \tag{10}$$

where $\tilde{A}$ is trivially related to A and we have used the definition given in (1). This proves our result.

GENERALIZATION TO MULTIPHOTON EMISSION

At the end of the second Section we have observed that the total decay amplitude for one photon emission off an external line is always proportional to the gauge-invariant function $I(A,B)$. In this section we consider the case of n soft-photons emitted from a transition involving N external charged lines. One might ask whether or not the radiative amplitude for the emission of an arbitrary number of photons is proportional to the product of the electromagnetic field tensors of each photon.

In the limit of very soft-photons, the dominant model-independent part of the radiative amplitude is given by [6]:

$$\mathcal{M}^{n} = \mathcal{M}_{0}\prod_{i=1}^{n}\sum_{j=1}^{N} e\eta_{j}\frac{p_{j}\cdot\varepsilon_{i}}{p_{j}\cdot k_{i}}\ . \tag{11}$$

In the above expression $\mathcal{M}_0$ denotes the corresponding non-radiative transition amplitude, ε_i is the polarization four-vector of the photon i and k_i its corresponding four-momentum ($k_i\cdot\varepsilon_i = 0$ and $k_i^2 = 0$, for real photons). The factor η_j denotes the electric charge of the $j-$th external charged particle in units of e. Note that when $\varepsilon_i = k_i$, the above result give rise to $\sum_{j=1}^{N} e\eta_j = 0$, which express the conservation of the electric charge in this process.

Note that the factors in eq. (11) can be written as follows:

$$\begin{aligned} \sum_{j=1}^{N} e\eta_j \frac{p_j \cdot \varepsilon_i}{p_j \cdot k_i} &= \sum_{j=1}^{N} e\eta_j \left(\frac{p_j \cdot \varepsilon_i}{p_j \cdot k_i} - \frac{P \cdot \varepsilon_i}{P \cdot k_i} \right) \\ &= \sum_{j=1}^{N} e\eta_j I_i(p_j, P) \,, \end{aligned} \tag{12}$$

where P is an arbitrary four-vector and the subindex i refers to the photon flag.

The above expression tell us that the model-independent part of the soft-photon amplitude can be also expressed in terms of the tensor for the electromagnetic field and is explicitly gauge-invariant. On the other hand, the model-dependent terms which are important for hard-photon emission, are gauge-invariant by themselves (no gauge cancellations occurs between different Feynman graphs), thus each one of them is proportional to $F_{\mu\nu}$.

The answer to this question is a straightforward generalization of the previous single photon case. We now consider a multiphoton amplitude process

$$\mathscr{M} = \varepsilon_1^{\mu} \varepsilon_2^{\nu} \cdots \varepsilon_n^{\omega} \mathscr{M}_{\mu\nu\cdots\omega} \,. \tag{13}$$

Gauge invariance requires that for each photon $k_i^{\delta} \mathscr{M}_{\mu\nu\cdots\delta\cdots\omega} = 0$.

Again, since the amplitude depends linearly on each photon polarization vector, $\mathscr{M}_{\mu\nu\cdots\omega}$ cannot depend on ε_i. For real photons, we have of course $\varepsilon_i \cdot k_i = k_i \cdot k_i = 0$, with the gauge or transversality conditions $\partial \cdot A_i = 0$ for each photon field .

The tensor $\mathscr{M}_{\mu\nu\cdots\delta\cdots\omega}$ can be expanded for each index in terms of any set of four linearly independent directions. We take two of them as the photon momentum k_i and polarization ε_i; the other independent four vectors will be denoted by a_i and b_i. Where each vector index is expanded in terms of its special basis: $\varepsilon_i, k_i, a_i, b_i$.

Any four vector index can be expanded in terms of this basis. Thus

$$\mathscr{M}_{\mu\nu\cdots\omega} = a_{i\delta} A_{i\bar{\delta}} + b_{i\delta} B_{i\bar{\delta}} + \varepsilon_{i\delta} C_{i\bar{\delta}} + k_{i\delta} D_{i\bar{\delta}}$$

where A, B, C and D are Lorentz covariant tensors in all but the δ index that corresponds to the i photon; we denote this by $\bar{\delta} \equiv \{\mu\nu \cdots \gamma\varepsilon \cdots \omega\}$.

The tensor C_i vanishes due to the independence of $\mathscr{M}_{\mu\nu\cdots\omega}$ upon ε_i. On the other hand, the conservation of electromagnetic current implies that for any photon

$$\begin{aligned} k_i \cdot \mathscr{M} &= k_i \cdot a A_i + k_i \cdot b_i B_i + k_i^2 D_i \\ &= 0 \,, \end{aligned} \tag{14}$$

thus, for on-shell photons we have

$$B_i = -\frac{k_i \cdot a_i}{k \cdot b_i} A_i \,,$$

We may thus conclude that

$$\mathscr{M}_{\mu\nu\cdots\omega} = (a_{i\delta} - b_{i\delta}\frac{k_i \cdot a_i}{k_i \cdot b_i})A_{i\bar{\delta}} + k_{i\delta}D_{i\bar{\delta}}$$

Once again the gauge (or more generally the transversality) condition $\varepsilon_i \cdot k_i = 0$ implies that the last term in the previous equation never contributes to the amplitude. Therefore

$$\mathscr{M}_{\mu\nu\cdots\omega} = (a_{i\delta} - b_{i\delta}\frac{k_i \cdot a_i}{k_i \cdot b_i})A_{i\bar{\delta}} \tag{15}$$

or

$$\begin{aligned} \mathscr{M} &= \varepsilon_1^{\mu} \cdots \varepsilon_n^{\omega} \mathscr{M}_{\mu\nu\cdots\omega} \\ &= \prod_i k_i \cdot a_i \, I(a_i, b_i) A \\ &= \prod_i F_i^{\mu_i\nu_i} a_{\mu_i} b_{\nu_i} \tilde{A} \, , \end{aligned} \tag{16}$$

where we have used the definition given in (1). Notice that one must still impose the symmetrization condition in all the photons which of course does not change our argument.

CONCLUSIONS

In this note we have shown that the physical amplitude of radiative processes depends linearly upon the field tensor $F_{\mu\nu}$ describing the electromagnetic field. This form of the amplitude is manifestly gauge-invariant. This factorization property can be generalized to space-time dimensions different from $3+1$.

It also allows a simple decompositon of the physical amplitude in terms of helicity states of the emitted photon in the form

$$\mathscr{F}_{\mu\nu}^{(\pm)} = \frac{1}{2}[F_{\mu\nu} \pm \widetilde{F}_{\mu\nu}] \, ,$$

namely,

$$\mathscr{M} = a_+^{\mu\nu} \mathscr{F}_{\mu\nu}^+ + a_-^{\mu\nu} \mathscr{F}_{\mu\nu}^- \, .$$

One of the practical uses of this factorization is that square of this amplitude is particularly simple when summed over photon polarizations [3].

ACKNOWLEDGMENTS

We would like to thank the organizers of this conference in honor of our friend Augusto García, who we will all miss. The assistance of Rodolfo Leo Méndez in producing the figures is greatly acknowledged. Finally, a critical reading of the manuscript by Dr. Jorge Flores is very much appreciated.

REFERENCES

1. A García, and P. Kielanowski, *The Beta Decay of Hyperons*, Lecture Notes in Physics **222**, Springer Verlag. Berlin-Heidelberg, 1985.
2. D.. N. Neville *Phys. Rev.*, **124**, 2037–2046 (1961).
3. S. G. Brown and S. A. Bludman *Phys. Rev.*, **136**, B1160–B1163 (1964).
4. G. López Castro and G. Toledo Sánchez *J. Phys.*, **G27**, 2203–2210 (2001).
5. G. López Castro and G. Toledo Sánchez, Phys. Rev. D **56**, 4408 (1997)
6. T. Muta, "Foundations of quantum chromodynamics", World Scientific Lecture Notes in Physics—Vol 5, Ch. 6, Singapore (1987)

Spontaneous Lorentz symmetry breaking in non-linear electrodynamics

Luis F. Urrutia [1]

Departamento de Física de Altas Energías, Instituto de Ciencias Nucleares, Universidad Nacional Autónoma de México, Apartado Postal 70-543, 04510 México D.F.

Abstract. A recently proposed model of non-linear electrodynamics arising from a gauge invariant spontaneous Lorentz symmetry breaking is reviewed. The potential providing the symmetry breaking is argued to arise from the integration of gauge bosons and fermions in an underlying theory. The invariant subgroups remaining after the symmetry breaking are determined, as well as the dispersion relations and polarization modes of the propagating linear sector or the model. Strong bounds upon the predicted anisotropy of the speed of light are obtained by embedding the model in the electromagnetic sector of the Standard Model Extension and taking advantage of the restrictions in the parameters derived there. Finally, a reasonable estimation of the intergalactic magnetic field is obtained by assuming that the vacuum energy of the model is described by the standard cosmological constant.

Keywords: Effective field theories. Spontaneous gauge invariant Lorentz symmetry breaking.
PACS: 11.30.Qc, 03.50.De, 11.10.Lm, 11.30.Cp

1. INTRODUCTION

Recent research in theories with spontaneous Lorentz symmetry breaking (SLSB) has focused in the discussion of photons arising as the corresponding Nambu-Goldstone bosons associated to such breaking. This is a very appealing proposal that intends to provide a dynamical setting for gauge invariance. This idea goes back to the works of Nambu [1] and Bjorken [2] together with many other contributions [3]. Recently it has been revived in references [4, 5, 6, 7, 8, 9]. A popular starting point to study this issue are the so called *bumblebee* models [10]. In the case of electrodynamics, one possible Lagrangian to consider is

$$L=-\frac{1}{4}B_{\mu\nu}B^{\mu\nu}-V(B_\mu B^\mu), \qquad B_{\mu\nu}=\partial_\mu B_\nu-\partial_\nu B_\mu, \tag{1}$$

where the potential term $V(B_\mu B^\mu)$ clearly violates gauge invariance. The Lorentz symmetry is subsequently broken by a non-zero vacuum expectation (VEV) value $< B^\mu >$, which minimizes the potential and which defines a preferred direction in space-time. This breaking splits the original four degrees of freedom in B^μ into three vectorial Nambu-Goldstone bosons A^μ , satisfying the constraint $A_\mu A^\mu = \pm M^2$, (to be identified with the photon) plus a massive scalar field σ, which is assumed to be exited only at very high energies. Under this threshold one basically recovers the Lagrangian for

[1] In memoriam Augusto García

CP1259, *First CINVESTAV-UNAM Symposium on High Energy Physics*
edited by J. C. D'Olivo, R. López-Fernández, and M. Á. Pérez

electrodynamics in vacuum plus the above mentioned non-linear constraint, which is interpreted as a gauge fixing condition. It is noteworthy to recall here that the constraint $A_\mu A^\mu = \pm M^2$ was originally proposed by Dirac as a way to derive the electromagnetic current from the additional excitations of the photon field, which now ceased to be gauge degrees of freedom, avoiding in this way the problems arising from considering point-like charges [11]. As reported in [12], this theory requires the existence of an ether emerging from quantum fluctuations, which does not necessarily leads to a violation of Lorentz invariance, due to a proposed averaging process of the quantum states producing such fluctuations.

The basic question to answer is to which extent the low energy limit of these theories describes the standard effects of QED, which have been experimentally measured with astonishing precision. Calculations at the tree level [1] and to the one loop level [8] have been carried on, showing that the possible Lorentz violating effects are not present in the physical observable quantities and that the results are completely consistent with the standard gauge invariant electrodynamics.

In this contribution, which summarizes the work done in collaboration with J. Alfaro in Ref. [13], we report a different approach, also based on SLSB, but preserving gauge invariance from the very beginning. To this end we choose to break the symmetry via a VEV of the gauge invariant electromagnetic tensor $F_{\mu\nu}$, instead of the vector potential A_μ. Two main challenges arise in this approach: (i) the first is to make plausible the existence of an effective potential $V(F)$ which can be argued to arise from a more fundamental theory, having an stable minimum to define the vacuum. (ii) The second challenge is to formulate the required non-linear electrodynamics in terms of the basic field $F_{\mu\nu}$ in such a way that we can introduce the standard vector potential only after the symmetry breaking is made in terms of the VEV $< F_{\mu\nu} >$.

2. THE MODEL

In relation to the issue (i) raised in the Introduction, the immediate possibility that comes to mind is to consider the effective photon interaction in QED after integrating the fermions . Unfortunately, a detailed calculation in the one and two-loop approximation produces, in the high constant magnetic field approximation,

$$V_{EM}(B) = \frac{1}{2}B^2 + V_{EM}^{(1)} + V_{EM}^{(2)} = \frac{1}{2}B^2 - \frac{e^2}{24\pi^2}B^2\left[\ln\left(\frac{eB}{m^2}\right)\right] - \frac{e^4}{128\pi^4}B^2\ln\left(\frac{eB}{m^2}\right), \quad (2)$$

which shows that the potential is unbounded from below [14].

An additional possibility arises in a further generalization of the mechanism proposed in Refs. [2, 15], restricted to the gauge invariant case. We start from a conventional gauge theory including fermions, gauge fields and Higgs fields which provide masses, via spontaneous symmetry breaking, to the gauge fields, except for the photon, which carries the electromagnetic $U(1)$ gauge invariance. For simplicity we consider only one fermion species, as in Ref.[2], but the idea can be generalized to many of them. In the latter case, the subsequent integration over the fermionic degrees of freedom can be made only for those having a mass larger than a given scale, thus leaving unintegrated the lightest ones.

Integrating the massive gauge bosons of this theory leads to a power series in bilinears of the fermion fields, which are suppressed by powers of the energy scale Λ that characterizes the SSB mechanism providing the mass to the gauge vector bosons [15]. We focus upon the neutral sector of the model where the following contributions to the effective Lagrangian arise

$$\frac{1}{\Lambda^{6n-4}}\left[\sum_a \left(\bar{\Psi}M_a\Psi\right)\left(\bar{\Psi}M^a\Psi\right)\right]^n, \ \frac{1}{\Lambda^{8n-4}}\left[\sum_a \left(\bar{\Psi}M_aD_\nu\Psi\right)\left(\bar{\Psi}M^aD^\nu\Psi\right)\right]^n, \dots \tag{3}$$

We are considering all possible contributions which are consistent with gauge invariance, the remaining symmetry of this sector of the theory. Notice also that terms containing the $U(1)$ covariant derivative $D_\nu = (\partial_\nu + ieA_\nu)$ are further suppressed with respect to those without it. Also we consider only the $n = 1$ case, which provides a gauge invariant generalization of the model in Ref. [2]. Thus our starting point is

$$L_{eff} = \bar{\Psi}\left(i\gamma^\mu\left(\partial_\mu + ieA_\mu\right) - m\right)\Psi - \sum_{M,a}\frac{r_M}{\Lambda^2}\left[\left(\bar{\Psi}M_a\Psi\right)\left(\bar{\Psi}M^a\Psi\right)\right]. \tag{4}$$

The index $M : \{S, V, T, PV, PS\}$ labels the tensorial objects that constitute the Dirac basis, that is to say Scalar, Vector, Tensor, Pseudovector and Pseudoscalar, respectively. The generic index a labels the covariant (contravariant) components of each tensor class. We follow the Dirac algebra conventions of Ref. [16], with appropriate factors chosen in such a way that each current $(j_M)_a = (\bar{\Psi}M_a\Psi)$ is real. In this way, the proposed generalization includes all possible fermionic quartic interactions. Also, the coefficients r_M are assumed to be of order one. The fermions can be integrated next by introducing the corresponding auxiliary fields $(C_M)_a$ for each real current $(j_M)_a$, as suggested in Ref. [15], and following standard manipulations in the path integral formulation [13]. In order to estimate the additional corrections provided by such currents to the purely electromagnetic effective potential (2) we choose to consider separately each of the contributions arising from the different currents and only add them to the total effective potential at the end. In this way we are neglecting possible interference terms among the currents, which nevertheless could be calculated within the approximations performed. Calling $V^{(2)}_{eff}$ the total effective potential arising from the currents j_M we obtain [13]

$$B \to \infty: \ V^{(2)}_{eff} = \left(\frac{1}{16\pi^2}\right)^2\left(\frac{m}{\Lambda}\right)^2 e^2B^2\left[\ln\left(\frac{2eB}{m^2}\right)\right]^2\sum_M r_M\,\theta_M, \tag{5}$$

$$B \to 0: \ V^{(2)}_{eff} = \left(\frac{1}{16\pi^2}\right)^2\left(\frac{m}{\Lambda}\right)^2\left(\frac{eB^2}{3m^2} + \frac{1}{18}\frac{e^2B^4}{m^4}\right)\sum_M r_M\,\theta_M. \tag{6}$$

Here m is the mass of the integrated fermion which is such that $m > \Lambda$. The main point of the estimation is that we obtain

$$\Theta = \sum_M r_M\,\theta_M = 4r_S + 16r_V + 48r_T - 16r_{PV} - 4r_{PS}, \tag{7}$$

which can be made positive from a judicious choice of the parameters r_M. In the high-intensity field approximation we observe that the dominant term in the electromagnetic contribution (2) can be overcame by the contribution from the additional currents provided $\Theta > 0$. It is important to remark that this preliminary estimation has only considered an appropriate regularization of the divergent integrals, but is missing an adequate renormalization of the total effective Lagrangian L_{eff}. Our renormalization conditions would be

$$\lim_{B\to 0} L_{eff} = 0, \quad \lim_{B\to 0}\left(\frac{dL_{eff}}{dB^2}\right) > 0, \tag{8}$$

in such a way that the effective potential is a decreasing function in the vicinity of $B = 0$. The final normalization to the value of $L_{eff} = +b^2/2$ will be imposed at the end of the procedure, after we have expanded the magnetic field around an stable minimum B_{Min} of V_{eff}, such that

$$B = B_{Min} + b \tag{9}$$

and we are left with the physical component b. This renormalization procedure is beyond the scope of these preliminary estimations.

Summarizing, we have made it plausible the existence of a gauge invariant potential $V_{eff}(F^2)$ having an stable minimum and arising as an effective low energy model from the integration of degrees of freedom corresponding to massive gauge-bosons and fermions in a more fundamental theory. This potential also exhibits the right behavior in the low-intensity field approximation. To examine the dynamical consequences of the symmetry breaking we have to adopt a convenient parametrization of such potential. There are various choices in the literature and we adopt the standard Ginzburg-Landau form

$$V(F_{\mu\nu}) = \frac{1}{2}\alpha F^2 + \frac{\beta}{4}\left(F^2\right)^2. \tag{10}$$

Next we settle the issue (ii) raised in the Introduction. Since our basic variable is $F_{\mu\nu}$, we start from the effective Lagrangian

$$L(F_{\alpha\beta}, X_\mu) = -V(F_{\alpha\beta}) - \bar{F}^{\nu\mu}\partial_\nu X_\mu, \quad F_{\alpha\beta} = -F_{\beta\alpha}, \tag{11}$$

where $\bar{F}^{\nu\mu}$ is the standard dual of the electromagnetic tensor. The fields X_μ are just Lagrange multipliers that will finally impose the condition that the field strength satisfies the $U(1)$ Bianchi identity and can then be expressed in terms of the vector potential. The above Lagrangian is similar to that used in the standard formulation of non-linear electrodynamics [17]. Our conventions are $\eta_{\mu\nu} = diag(+,-,-,-), \quad \varepsilon^{0123} = +1, \quad \varepsilon_{123} = +1$.

In order to find the vacuum configuration we have to extremize the effective action, subjected to the condition that $F_{\alpha\beta}$ and X_α are constant fields in order to preserve translational invariance. Applying these requirements to (11) plus the choice (10) we obtain the vacuum condition

$$\left(F_{\mu\nu}\right)_V = C_{\mu\nu}, \quad \left(F^2\right)_V = -\frac{\alpha}{\beta} = C^2 \neq 0. \tag{12}$$

The expansion around the minimum $(C_{\mu\nu},\, C_\mu)$ is written as

$$F_{\alpha\beta}(x) = C_{\alpha\beta} + a_{\alpha\beta}(x), \qquad X_\mu = C_\mu + \bar{X}_\mu, \tag{13}$$

and leads to

$$V(F_{\alpha\beta}) = V(C_{\alpha\beta}) + \beta\left(C_{\alpha\beta}a^{\alpha\gamma}\right)^2 + \beta\left(C_{\alpha\beta}a^{\alpha\beta}\right)\left(a_{\mu\nu}a^{\mu\nu}\right) + \frac{\beta}{4}\left(a_{\alpha\beta}a^{\alpha\beta}\right)^2 \tag{14}$$

$$\equiv V(C_{\alpha\beta}) + \bar{V}(a_{\alpha\beta}). \tag{15}$$

We notice that (14) has no linear term in $a^{\alpha\gamma}$, and also that the quadratic term $\beta\left(C_{\alpha\beta}a^{\alpha\gamma}\right)^2$ is positive provided $\beta > 0$, thus verifying the expansion around a minimum.

In this way, the spontaneously symmetry broken action is

$$S(a_{\alpha\beta}, \bar{X}_\mu) = -\int d^4x \left(\frac{1}{2}\varepsilon^{\nu\mu\alpha\beta}a_{\alpha\beta}\partial_\nu\bar{X}_\mu + \bar{V}(a_{\alpha\beta}) + V(C)\right), \tag{16}$$

where we have eliminated the total derivatives arising from the shifts of the fields (13) performed in the original Lagrangian (11). From the equations of motion arising from (16) one can show that the Lagrange multiplier $\bar{X}_\mu$ is fully determined up to gauge transformation $\bar{X}_\mu \longrightarrow \bar{X}_\mu + \partial_\mu\chi$. Also, the equation of motion derived from the variation $\delta\bar{X}_\mu$ establishes that the two form a satisfies $da = 0$. The Hodge-De Rham theorem says that the most general solution to this is obtained by requiring $a = dA + ls$, where A is a one form, l is a constant and s is an harmonic two-form. Since in our case we naturally have one such form at our disposal, arising precisely from the chosen vacuum, we take $s_{\mu\nu} = C_{\mu\nu}$. Calling $dA = f$ we have

$$a_{\alpha\beta}(x) = lC_{\alpha\beta} + f_{\alpha\beta}(x). \tag{17}$$

Substituting this solution in (16) we are left with the reduced effective action

$$S(A_\alpha) = -\int d^4x\, \left(\bar{V}(a_{\alpha\beta}) + V(C)\right), \qquad a_{\alpha\beta}(x) = lC_{\alpha\beta} + \partial_\alpha A_\beta - \partial_\beta A_\alpha. \tag{18}$$

After a convenient redefinition of parameters ($\beta \to \mathscr{B}, \quad C_{\mu\nu} \to D_{\mu\nu}$) we obtain our final effective action

$$S(A_\alpha) = \int d^4x \left(-\frac{1}{4}\left[1 - D^2\mathscr{B}\right]D^2 - \frac{1}{4}f_{\mu\nu}f^{\mu\nu} - \mathscr{B}\left[\left(D_{\mu\nu}f^{\mu\nu}\right) + \left(f_{\mu\nu}f^{\mu\nu}\right)\right]^2\right). \tag{19}$$

The only restriction upon the parameters is $\mathscr{B} > 0$, with D^2 arbitrary.

3. SYMMETRY ALGEBRAS ARISING FROM DIFFERENT CHOICES OF THE VACUUM

In this section we study the possible vacua allowed by the tensor symmetry breaking and we also identify the corresponding subgroups of the Lorentz group which are left invariant after the breaking. In order to do this, it is convenient to parameterize the background field $D_{\mu\nu}$ in terms of its three dimensional components **e** and **b**. Since we have two vectors that determine a plane, we choose a coordinate frame such that

$$\mathbf{b} = (0, 0, b), \qquad \mathbf{e} = (0, \quad e_2 = |\mathbf{e}|\sin\psi, \quad e_3 = |\mathbf{e}|\cos\psi), \tag{20}$$

In this way the matrix representing the vacuum is

$$[D_{\mu\nu}] = \begin{bmatrix} 0 & 0 & e_2 & e_3 \\ 0 & 0 & -b & 0 \\ -e_2 & b & 0 & 0 \\ -e_3 & 0 & 0 & 0 \end{bmatrix}. \tag{21}$$

The most general infinitesimal generator G, including Lorentz transformations plus dilatations is

$$G = [G^{\mu}{}_{\nu}] = -i \begin{bmatrix} z & x_1 & x_2 & x_3 \\ x_1 & z & -y_3 & y_2 \\ x_2 & y_3 & z & -y_1 \\ x_3 & -y_2 & y_1 & z \end{bmatrix}. \tag{22}$$

Motivated by the work in Ref. [18] we are including conformal dilatations D among our generators. Within this restricted Poincare algebra, this generator commutes with the remaining ones corresponding to pure Lorentz transformations and can be realized as a multiple of the identity. We do this in order to explore the possibility of having the largest possible invariant sub-algebra after the symmetry breaking. The condition for the vacuum to be invariant under the infinitesimal transformations is

$$0 = G^{\mu}{}_{\alpha} D^{\alpha\nu} + G^{\nu}_{\alpha} D^{\mu\alpha} \tag{23}$$

and the resulting equations are

$$x_2 b - y_2 e_3 = -y_3 e_2, \qquad x_2 e_3 + y_2 b = x_3 e_2, \tag{24}$$
$$x_1 e_2 + 2zb = 0, \qquad y_1 e_2 + 2z e_3 = 0, \tag{25}$$
$$-x_1 b + y_1 e_3 = 2z e_2, \qquad x_1 e_3 + y_1 b = 0. \tag{26}$$

By studying the solutions of the above system of equations corresponding to the seven non-trivial possibilities dictated by the choices in the arrangement $(b,\ e_2,\ e_3)$ in the above coordinate system, we conclude that the only remaining invariant subgroups correspond to various two-parameter subalgebras which are isomorphic to $T(2)$ and only one three-parameter subalgebra which is isomorphic to $HOM(2)$.

4. EQUATIONS OF MOTION, DISPERSION RELATIONS AND POLARIZATIONS

To study the propagation properties we consider only the quadratic terms in the effective Lagrangian

$$L_0 = -\frac{1}{4} f_{\mu\nu} f^{\mu\nu} - \mathscr{B} \left(f_{\mu\nu} D^{\mu\nu} \right)^2, \tag{27}$$

The equations are

$$\left(\partial^2 A_\beta - \partial_\beta \partial^\alpha A_\alpha \right) = -8 \mathscr{B} D_{\alpha\beta} \partial^\alpha \left(D^{\mu\nu} \partial_\mu A_\nu \right). \tag{28}$$

We search for plane wave solutions of equations (28) having a space-time dependence of any field being proportional to $e^{i(\mathbf{k}\cdot\mathbf{x}-\omega t)}$, where $k_\alpha = (\omega, \mathbf{k})$ is the four-momentum of the wave. We choose the Coulomb gauge and split (28) into $0+3$ contributions. The modified photon dispersion relations together with the corresponding plane wave polarizations are classified according to the values of $D_{i\alpha}k^\alpha A^i$. The amplitudes of the propagating fields are written in terms of those parameterizing $D_{\mu\nu}$. The case $D_{i\alpha}k^\alpha A^i = 0$ leads to dispersion relations $\omega = |\mathbf{k}|$ with the triad $\mathbf{E}$, $\mathbf{B}$, $\mathbf{k}$ having standard orthogonality properties. The situation $D_{i\alpha}k^\alpha A^i \neq 0$ produces the following properties for the vectors involved: $\mathbf{E}$ is perpendicular to $\mathbf{B}$, $\mathbf{B}$ is perpendicular to $\mathbf{k}$, but $\mathbf{E}$ is not necessarily orthogonal to $\mathbf{k}$, The corresponding dispersion relations are of the form $\omega = |\mathbf{k}|\, F(\text{angles})$, where the function F is independent of $|\mathbf{k}|$, been only function of the angles between $\mathbf{k}$ and $\mathbf{e}$, $\mathbf{k}$ and $\mathbf{b}$, $\mathbf{k}$ and $\mathbf{e}\times\mathbf{b}$. In this way our model predicts anisotropy in the speed of light. To the leading order $\mathscr{B}|\mathbf{e}|^2$, $\mathscr{B}|\mathbf{b}|^2$, $\mathscr{B}|\mathbf{e}||\mathbf{b}|$ the magnitude of such velocity is

$$c(\hat{\mathbf{k}}) = 1 + 8\mathscr{B}\left(e^2+b^2\right) - 4\mathscr{B}\left(\left(\mathbf{b}\cdot\hat{\mathbf{k}}\right)^2 + \left(\mathbf{e}\cdot\hat{\mathbf{k}}\right)^2 - 2\hat{\mathbf{k}}\cdot\left(\mathbf{e}\times\mathbf{b}\right)\right). \qquad (29)$$

Let us emphasize that the above result guarantees the stability of the model in the small LIV approximation.

5. THE MODEL AS A SECTOR OF THE SME

In order to impose some bounds upon the parameters of the model it is convenient to recast the quadratic sector of the action (19) in the language of the Standard Model Extension [19] and to make use of the numerous experimental constraints derived from it (see for example Ref. [20]) and summarized in Table II: Photon-sector summary, of Ref. [21]. To this end we make the identification

$$-\mathscr{B}\left(f_{\mu\nu}D^{\mu\nu}\right)^2 = -\frac{1}{4}\left(k_F\right)^{\kappa\lambda\mu\nu} f_{\kappa\lambda} f_{\mu\nu}, \qquad (30)$$

and recast $(k_F)^{\kappa\lambda\mu\nu}$ in terms of $D_{\mu\nu}$ obtaining

$$\left(k_F\right)^{\kappa\lambda\mu\nu} = 4\mathscr{B}\left[D^{\kappa\lambda}D^{\mu\nu} + \frac{1}{2}\left(D^{\kappa\mu}D^{\lambda\nu} - D^{\lambda\mu}D^{\kappa\nu}\right)\right] - \frac{1}{2}\mathscr{B}D^2\left(\eta^{\kappa\mu}\eta^{\lambda\nu} - \eta^{\lambda\mu}\eta^{\kappa\nu}\right), \qquad (31)$$

where $D^2 = 2(\mathbf{b}^2 - \mathbf{e}^2)$. The linear combinations of the components $(k_F)^{\kappa\lambda\mu\nu}$ in terms of which the bounds are presented in Ref. [21] have been defined in Ref. [20]. Their expression in terms of our variables, obtained from the relation in Eq. (31), are

$$\left(\bar{\kappa}_{e+}\right)^{jk} = 6\mathscr{B}\left[b_j b_k - e_j e_k - \frac{1}{3}\left(\mathbf{b}^2 - \mathbf{e}^2\right)\delta_{jk}\right] < 10^{-32}, \qquad (32)$$

$$\left(\bar{\kappa}_{e-}\right)^{jk} = -6\mathscr{B}\left[\left(b_j b_k + e_j e_k\right) - \frac{1}{3}\delta_{jk}\left(\mathbf{b}^2 + \mathbf{e}^2\right)\right] < 10^{-16}, \qquad (33)$$

$$\left(\bar{\kappa}_{o+}\right)^{jk} = 6\mathscr{B}\left(e_j b_k - e_k b_j\right) < 10^{-12}, \qquad (34)$$

$$(\bar{\kappa}_{o-})^{jk} = 6\mathscr{B}\left[(e_j b_k + e_k b_j) - \frac{2}{3}(\mathbf{e}\cdot\mathbf{b})\,\delta_{jk}\right] < 10^{-32}, \tag{35}$$

$$\bar{\kappa}_{tr} = -2\mathscr{B}(\mathbf{e}^2 + \mathbf{b}^2) < 10^{-15}. \tag{36}$$

The bounds written at the end of each of the above equations are valid in the Standard Inertial Reference Frame centered in the Sun defined in Ref. [20]. The bounds for $\bar{\kappa}_{tr}$ have recently been improved from 10^{-7} [21], to 10^{-11} [22], and finally to 10^{-15} [23]. In quoting our results arising from the bounds upon $(\bar{\kappa}_{e+})^{jk}$, $(\bar{\kappa}_{o-})^{jk}$ and $\bar{\kappa}_{tr}$ we have chosen our reference frame in such a way that the light-cone gets only modified by the photon LIV parameters, while the propagation of the fermions is the standard one [20],[24]. Notice that the bilinears which are odd under the duality transformations $\mathbf{e} \to \mathbf{b}$, $\mathbf{b} \to -\mathbf{e}$, are much more suppressed than those which are even. In this way, even though our model is not duality invariant, this transformation seems to explain the above mentioned hierarchy in the LIV parameters exhibited in Eqs. (32)-(36). A careful study of Eqs. (32)-(36) leads to the restriction

$$\mathscr{B}(\mathbf{e}^2 + \mathbf{b}^2) = \frac{1}{M_{\mathscr{B}}^4}(\mathbf{e}^2 + \mathbf{b}^2) < 2.5 \times 10^{-33}, \tag{37}$$

which fulfills all the constraints separately established in them.

6. SUMMARY AND FINAL COMMENTS

We have presented a novel way of implementing a model with spontaneously broken Lorentz symmetry by introducing a constant VEV of the field strength $\langle F_{\mu\nu}\rangle = C_{\mu\nu}$. We start from the effective Lagrangian (11) describing a non-linear electrodynamics and containing a potential $V_{eff}(F_{\mu\nu})$, which is argued to arise after integrating massive gauge bosons and fermions in an underlying conventional theory. This approach is an extension of the model in Refs. [2, 15], which includes all quartic fermion interactions allowed by the Dirac algebra. For simplicity we have considered only one fermion species. The subsequent integration of them produces the effective gauge invariant potential which is bounded in the high-intensity field limit and has an stable minimum. In order to investigate the dynamical consequences of the electrodynamics constructed around such minimum we have chosen the standard Ginzburg-Landau parametrization for $V_{eff}(F_{\mu\nu})$. We have shown that the Lorentz subgroups surviving after the breaking are only $T(2)$ and $HOM(2)$. The modified photon dispersion relations together with the corresponding plane wave polarizations have been classified in Section 4..

In order to make a more quantitative statement about the predicted anisotropy of the speed of light, let us start from Eq. (29) and define the corresponding two-way speed of light

$$c_{TW}(\hat{\mathbf{k}}) = \frac{1}{2}\left(c(\hat{\mathbf{k}}) + c(-\hat{\mathbf{k}})\right) = 1 + 8\mathscr{B}\left(e^2 + b^2\right) - 4\mathscr{B}\left(\left(\mathbf{b}\cdot\hat{\mathbf{k}}\right)^2 + \left(\mathbf{e}\cdot\hat{\mathbf{k}}\right)^2\right). \tag{38}$$

A standard measure for the anisotropy of the speed of light is the comparison of the two-way velocities in perpendicular directions. We single out the vector $\left(\hat{\mathbf{e}} \times \hat{\mathbf{b}}\right) = \hat{\mathbf{n}}$ in our reference frame. The vectors $\hat{\mathbf{k}}$ and $\hat{\mathbf{n}}$ form a plane in which we define the vector $\hat{\mathbf{q}}$ perpendicular to $\hat{\mathbf{k}}$ given by $\hat{\mathbf{q}} = \hat{\mathbf{k}} \times \left(\hat{\mathbf{k}} \times \hat{\mathbf{n}}\right) = \left(\hat{\mathbf{k}} \cdot \hat{\mathbf{n}}\right)\hat{\mathbf{k}} - \hat{\mathbf{n}}$. Next we calculate the two-way light speed along this perpendicular direction obtaining

$$c_{TW}(\hat{\mathbf{q}}) = 1 + 8\mathscr{B}\left(e^2 + b^2\right) - 4\mathscr{B}\left(\hat{\mathbf{k}} \cdot \hat{\mathbf{n}}\right)^2 \left(b^2\left(\hat{\mathbf{b}} \cdot \hat{\mathbf{k}}\right)^2 + e^2\left(\hat{\mathbf{e}} \cdot \hat{\mathbf{k}}\right)^2\right). \tag{39}$$

An appropriate definition for the anisotropy of the speed of light in this model is

$$\frac{\Delta c}{c} \equiv \left|c_{TW}(\hat{\mathbf{k}}) - c_{TW}(\hat{\mathbf{q}})\right| = \left|4\mathscr{B}\left(1 - \left(\hat{\mathbf{k}} \cdot \hat{\mathbf{n}}\right)^2\right)\left(b^2\left(\hat{\mathbf{b}} \cdot \hat{\mathbf{k}}\right)^2 + e^2\left(\hat{\mathbf{e}} \cdot \hat{\mathbf{k}}\right)^2\right)\right|. \tag{40}$$

From the last expression we obtain the bound

$$\frac{\Delta c}{c} < 4\mathscr{B}\sin^2\theta(\mathbf{b}^2 + \mathbf{e}^2) < 4\mathscr{B}(\mathbf{b}^2 + \mathbf{e}^2) < 10^{-32}, \tag{41}$$

according to Eq. (37), where θ is the angle between the vector $\hat{\mathbf{n}}$ and $\hat{\mathbf{k}}$. The standard Michelson-Morley type of analysis can be made by measuring first the time difference along two perpendicular trajectories L_1 (propagating with $c_{TW}(\hat{\mathbf{k}})$), and L_2 (propagating with $c_{TW}(\hat{\mathbf{q}})$)

$$\tau = \frac{L_2}{c_{TW}(\hat{\mathbf{q}})} - \frac{L_1}{c_{TW}(\hat{\mathbf{k}})}. \tag{42}$$

Subsequently a similar measurement is made by rotating the apparatus by 90 degrees

$$\tau' = \frac{L_2}{c_{TW}(-\hat{\mathbf{k}})} - \frac{L_1}{c_{TW}(\hat{\mathbf{q}})}. \tag{43}$$

Finally, the difference in optical paths

$$\Delta\tau = \left|\tau - \tau'\right| = 4\mathscr{B}(L_2 + L_1)\sin^2\theta\left(b^2\left(\hat{\mathbf{b}} \cdot \hat{\mathbf{k}}\right)^2 + e^2\left(\hat{\mathbf{e}} \cdot \hat{\mathbf{k}}\right)^2\right) \tag{44}$$

measures the change in the interference pattern of the Michelson-Morley experiment, according to our model. Perhaps a more direct comparison of our results can be made with the work of Ref. [25], where the Euler model of nonlinear electrodynamics [26] is used to predict the following bound on the isotropic component of the one-way speed of light

$$\frac{\tilde{\delta}c}{c} \leq 1.2 \times 10^{-23}. \tag{45}$$

In our case this would correspond to averaging over all angles in (29), leading to

$$\frac{\tilde{\delta}c}{c} \leq 8\mathscr{B}\left(e^2 + b^2\right) \simeq 2 \times 10^{-32}. \tag{46}$$

Let us emphasize that in all our estimations we are assuming that the involved references frames (CMB, Sun based, earth based) are concordant, in such a way that first order quantities in the Lorentz invariance violation parameters remain of the same order in all of them.

Assuming that our background fields **e** and **b** might represent some galactic or intergalactic fields in the actual era, we obtain a very reasonable bound for the magnetic intergalactic field by assuming that the constant appearing in the action (19) corresponds to an energy density ρ

$$\rho = \frac{1}{4}\left[(1 - D^2\mathscr{B}\right] D^2 \simeq \frac{1}{2}\left(\mathbf{b}^2 - \mathbf{e}^2\right), \tag{47}$$

which we can associate with the cosmological constant, since it would represent a global property of the universe. The fact that this constant is positive favors $\left(\mathbf{b}^2 - \mathbf{e}^2\right) > 0$, so that one can perform a passive Lorentz transformation to a reference frame where $\mathbf{e} = \mathbf{0}$. Supposing further that this frame, which describes the intergalactic fields, is concordant with the Standard Inertial Reference Frame and taking the upper limit [27]

$$|\rho_\Lambda| < 10^{-48}\ (GeV)^4, \tag{48}$$

we obtain the bound

$$|\mathbf{b}|\ < 5 \times 10^{-5}\ Gauss, \tag{49}$$

which is consistent with observations of intergalactic magnetic fields [28]. Let us recall that $1\ Gauss = 1.95 \times 10^{-20}\ (GeV)^2$. It is amusing to observe that this bound on $|\mathbf{b}|$ translates into the following condition for the mass scale M_B introduced in Eq. (37)

$$M_B > 1.4 \times 10^{-4}\ GeV. \tag{50}$$

Recalling that the lightest charged particle is the electron with $m_e = 5.1 \times 10^{-4}\ GeV$, we notice that the above bound is consistent with the QED Euler-Heisenberg type of non-linear corrections to the photon interaction, which scale as $1/m^4$, with m being the mass of the charged particle contributing to the loop in the effective action [29].

ACKNOWLEDGMENTS

L.F.U is partially supported by projects CONACYT-55310 and DGAPA-UNAM-IN111210.

REFERENCES

1. Y. Nambu, Suppl. of the Prog. Theor. Phys., Extra Number, 190 (1968).
2. J. D. Bjorken, Ann. Phys. (N.Y.) **24,** 174(1963); *Emergent Gauge Bosons*, Contributed to 4th Workshop on What Comes Beyond the Standard Model?, Bled, Slovenia, 17-27 Jul 2001, arXiv: hep-th/0111196.
3. G. S. Guralnik, Phys. Rev. **134**, B 1404(1964); A. Kovner abd B. Rosenstein, Phys. Rev. **D49**, 5571(1994); I. Low and A. V. Manohar, Phys. Rev. Lett. **88**, 101602(2002).
4. V.A. Kostelecky and R. Potting, Gen. Rel. Grav. **37**, 1675(2005), Int. J. Mod. Phys. **D14**, 2341(2005).
5. J.L. Chkareuli, C.D. Froggatt and H.B. Nielsen, Nucl. Phys. **B821**, 65 (2009).

6. A. T. Azatov and J.L. Chkareuli, Phys. Rev. **D73**, 065026(2006).
7. J.L. Chkareuli,C.D. Froggatt, R. N. Mohapatra and H.B. Nielsen, *Photon as a Vector Goldstone Boson: Nonlinear σ Model for QED*, arXiv: hep-th/0412225.
8. J.L. Chkareuli and J. G. Jejelava, Phys. Lett. **B659**, 754(2008).
9. V. A. Kostelecky and R. Potting, Phys. Rev. **D79**, 0065018 (2009).
10. R. Bluhm and V. A. Kostelecky, Phys. Rev. **D71**, 065008(2005).
11. P.A.M. Dirac, Proc. Roy. Soc. (London), **A209**, 291(1951); P.A.M. Dirac, Proc. Roy. Soc. (London), **A212**, 330(1951); P. A. M. Dirac, Proc. Roy. Soc. (London), **A223**, 438(1954).
12. P.A.M. Dirac, Nature (London) **168**, 906 (1951).
13. J. Alfaro and L. F. Urrutia, Phys. Rev. **D81**, 025007 (2010).
14. See for example: W. Dittrich and M. Reuter, *Effective Lagrangians in Quantum Electrodynamics,* Lecture Notes in Physics 220, Springer Verlag, Berlin, 1985.
15. P. Kraus and E. T. Tomboulis, Phys. Rev. **D66**, 045015 (2002).
16. J.D. Bjorken and S.D. Drell, *Relativistic Quantum Mechanics*, McGraw-Hill Book Company, New York, 1964.
17. J. Plebanski, *Lectures on Non-Linear Electrodynamics*, Nordita, 1970.
18. A. J. Hariton and R. Lehnert, Phys. Lett. **A367**, 11(2007).
19. D. Colladay and V. A. Kostelecky, Phys. Rev. **D55**, 6760 (1997); **D58**, 116002 (1998).
20. V. A. Kostelecky and M. Mewes, Phys. Rev. **D66**, 056005(1979).
21. V. A. Kostelecky and N. Russell, *Data Tables for Lorentz and CPT Violation* , arXiv: 0801:0287 [hep-ph].
22. M. A. Hohensee, R. Lehnert, D. F. Phillips and R. L. Walsworth, Phys. Rev. Lett. **102,** 170402 (2009).
23. B. Altschul, Phys. Rev. **D80**, 091901 (2009).
24. Q.G. Bailey and V. A. Kostelecky, Phys. Rev. **D70**, 076006 (2004). R. Lehnert, private comunication.
25. A. M. Grassi-Strini, G. Strini and G. Tagliaferri, Phys. Rev. **D19**, 2330(1979).
26. H. Euler, Ann. Phys. **26**, 398 (1936).
27. S. M. Carrol, Liv. Rev. Rel. **4**: 1, (2001), arXiv: astro-ph/00004075
28. See for example K. H. Radler, in *The Generation of Cosmic Magnetic Fields*, Lecture Notes in Physics # 556, edited by D. Page and J. Hirsch, Springer-Verlag, Berlin, 2000.
29. See for example W. Greiner and J. Reinhardt, *Quantum Electrodynamics*, Third Edition, Springer-Verlag, Berlin, 2003.

Author Index